INTERMEDIATE ALGEBRA

With

Analytic Geometry

Alice Gorguis

To order additional copies of this book, contact:
Xlibris
1-888-795-4274
www.Xlibris.com
Orders@Xlibris.com

Dedicated to My Students of Intermediate Algebra

Contents

Preface

This book is designed for students who need additional preparation for math classes, at North Park University, numbered 1020 or higher. The book covers all the topics that are required by the school which are: topics in beginning and intermediate algebra such as: equations and inequalities, systems, polynomials, factoring, graphing, roots and radicals, rational functions, quadratic equations, and conic sections. Most of the topics covered in this book are applied in our daily life; the proof is in the application sections. For some students this is the last course they need to take, for others just the first in series of many depending on students' major.
The text explains the method of solving problems step by step, and shows more than one method to solve a problem, also shows how to solve problems graphically and using technology (TI) so the student will learn how to solve the problem algebraically, then see the solution graphically. Most of the chapters are ended with application section to learn how to apply the topic to real life situation. Also, at the end of each chapter there is a chapter exercise, and a chapter test, students can use the chapter exercise as a study guide for test
My goal of writing this book is to make it easy for students to read through each page and doesn't overwhelm, or complicate Intermediate Algebra, but gives the skills and practice that they need.

Best Wishes
Alice Gorguis

1. Linear Equations

1. Linear Equations

Objectives: 1. Introduction

2. Linear Equations and Solutions

3. Applications of Linear Equations

4. Exponents and their properties

5. Scientific Notations

1.1 Introduction

Definition: Equation is a mathematical statement contains two expressions that are separated by the equal sign. The equation contains one or more variables (unknowns), also can be expressed as a balanced scale.

Type of equations: equation can be linear or nonlinear.

Linear equations: is the one with variable (un-known) of power=1.

Non-linear equation: is the one with unknown power > 1.

Solving equation: means to find the numerical value for the unknown.

Solution: means the value of the unknown that makes the equation a true statement.

Solution set: is the set that contains all the values that makes the equation a true statement. { } is the set symbol, or empty set. Ø is the symbol for empty set too.

1.2 Linear Equations and Solutions

Objectives: Solving Linear Equation in One Variable
- Algebraically
- Graphically

A linear equation in one variable is the equation of the following form:
$$ax + b = 0 \dots (1)$$
Where, a, and b are real numbers, and (a) is the coefficient of the unknown (variable) x, $a \neq 0$. Equation (1) has exactly one solution, and can be solved as follows:

1. Subtract b from both sides: $ax + \cancel{b} = 0$
$$\cancel{-b} - b$$
$$\overline{\rule{3cm}{0.4pt}}$$
$$ax = -b$$

2. Divide both sides by (a) the coefficient of x gives → $x = -b/a$

Example-1	Solve the equation: $5x - 2 = 13$

Algebraic Solution $5x - \cancel{2} = 13$ Add 2 on both sides
$$+\cancel{2} \quad +2$$
$$\overline{\rule{3cm}{0.4pt}}$$
$$\cancel{5}x \quad = 15 \quad \text{Divide both sides by 5 then} \rightarrow x = 3$$
$$\cancel{5} \qquad 5$$

To check the solution, we replace the unknown x with its calculated value 3, and see if the

R.S = L.S as follows: ?
$$5(3) - 2 = 13$$
$$15 - 2 = 13$$
$$13 = 13 \text{ this is a true statement, then x=3 is the true solution}$$

Graphical Solution: Using TI-83
Enter the left side on Y1 = 5x – 2
Enter the right side on Y2= 13
Click on Graph to get:

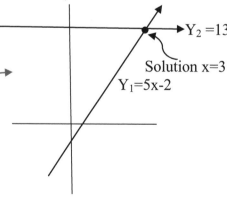

$Y_2 = 13$

Solution x=3

$Y_1 = 5x-2$

On TI-83 you'll find the solution under 2nd Table

X	Y1	Y2
0	-2	13
2	8	13
3	13	13
4	18	13

Practice-1: Solve and Check
$2x–2=10$

Example-2	Solve the equation; $-7x + 12 = 3 - 2x$

Solution

$-7x + 12 = 3 - 2x$ Group the terms with variable x on left side
$\underline{+2x \qquad\qquad + 2x}$

$-5x + 12 = 3$ Group the constants on right side
$\underline{\qquad - 12 \ -12}$

$-5x \qquad = -9$

$\dfrac{-5x}{-5} \qquad = \dfrac{-9}{-5}$ divide both sides by (-5) the coefficient of x:

Then → $x = \dfrac{9}{5} = 1.8$

Check: replace x with the value 9/5 as follows:

$-7(9/5) + 12 = 3 - 2(9/5)$ multiply both sides by 5

$5\{-7(9/5) + 12 = 3 - 2(9/5)\}$

$-7(9) + 5.12 = 5. \, 3 - 2(9)$

$-63 + 60 = 15 - 18$

$-3 = -3$ True statement, then→ the true solution is x = 9/5 = 1.8 or the solution set is {9/5}

Graphical Solution: Using TI-83

Enter the left side on Y1 = –7x + 12

Enter the right side on Y2= 3 – 2x

X	Y1	Y2
1.7	.1	-.4
1.8	-.6	-.6
1.9	-1.3	-.8
2	-2	-1

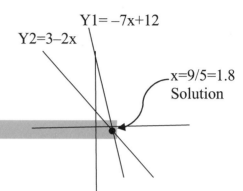

Y1= –7x+12

Y2=3–2x

x=9/5=1.8
Solution

Practice-2: Solve and Check

$-9x + 15 = 4 - 3x$

Example-3	Solve the equation: $x - 8 = \dfrac{12}{5}$

Solution

$$x - 8 = \frac{12}{5}$$

$$+8 \qquad +8 \qquad \text{Add 8 to both sides}$$

$$x \quad = \frac{12}{5} + 8$$

$$x \quad = \frac{12}{5} + \frac{8.5}{1.5}$$

$$x = \frac{12}{5} + \frac{40}{5} \quad \text{common LCD} = 5$$

$$x = \frac{12 + 40}{5} = \frac{52}{5} \rightarrow \text{then } x = \{52/5\} \text{ is the solution}$$

Check: Replace x with $\dfrac{52}{5}$

Left side (L.S) is: $\dfrac{52}{5} - 8 = \dfrac{52}{5} - \dfrac{8.5}{1.5}$

$$= \frac{52}{5} - \frac{40}{5} \quad \text{Common LCD} = 5$$

$$\text{The left side (L.S)} = \frac{52 - 40}{5} = \frac{12}{5} = \text{right side (R.S)}$$

Graphical Solution: Using TI-83
Enter the left side on Y1 = x–8
Enter the right side on Y2= 12/5

X	Y1	Y2
10.3	2.3	2.4
10.4	2.4	2.4
10.5	2.5	2.4

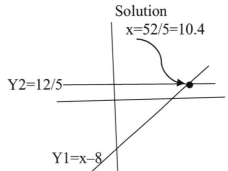

> **Practice-3**: Solve and Check
>
> $$x-6 = \frac{10}{3}$$

Example-4	Solve the equation: $5x - 8 + 2x = 15 + 6x - 3$

Solution $5x - 8 + 2x = 15 + 6x - 3$

Simplify on each side first by grouping the terms with variable x next to each other, and constants terms next to each other as follows:

$$5x + 2x - 8 = 15 - 3 + 6x$$

$$7x - 8 = 12 + 6x$$

Then group constant terms from both sides next to each other, and terms with x variable next to each other:

$7x - 6x = 12 + 8$. Then → $x = 20$ is the solution

Check: replace each x with 20:
$$5(20) - 8 + 2(20) = 15 + 6(20) - 3$$
$$100 - 8 + 40 = 15 + 120 - 3$$
$$132 = 132$$

Graphical Solution: Using TI-83
Enter the left side on Y1 = 7x – 8
Enter the right side on Y2 = 12 + 6x

X	Y1	Y2
19.9	131.3	131.4
20	132	132
20.1	132.7	132.6

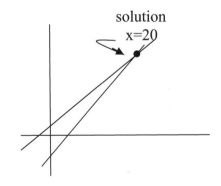

solution
x=20

Practice-4: Solve and Check

$$9x - 11 + 3x = 22 + 6x - 3$$

| Example-5 | Solve the equation: $3(5x - 8) + 2x = 2(6x - 3)$ |

Solution: $3(5x - 8) + 2x = 2(6x - 3)$

Simplify the expressions on each side, using distribution,

$$3(5x - 8) + 2x = 2(6x - 3)$$

$3.\ 5x - 3.8 + 2x = 2.\ 6x - 2.3$

$15x - 24 + 2x = 12x - 6$

$15x + 2x - 24 = 12x - 6$

$17x - 24 = 12x - 6$
$\underline{-12x \qquad\quad -12x}$

$5x - 24 = \qquad -6$
$\underline{\quad +24 \qquad\quad +24}$

$\dfrac{\cancel{5}x}{\cancel{5}} \quad = \quad \dfrac{18}{5} \qquad$ Then $\rightarrow x = \dfrac{18}{5}$

Check: Replace x with $\dfrac{18}{5}$

$$3\ (5.\ \frac{18}{5} - 8) + 2.\ \frac{18}{5} = 2\ (6.\ \frac{18}{5} - 3)$$

Open the parentheses and use distribution:

$$3.\ \frac{18}{\cancel{5}} - 3.\ 8 + 2.\ \frac{18}{5} = 2.\ 6.\ \frac{18}{5} - 2.\ 3$$

$$3.\ 18 - 24 + \frac{2.18}{5} = \frac{12.18}{5} - 6$$

$$54 - 24 + \frac{36}{5} = \frac{216}{5} - 6$$

Multiply both sides by the common LCD=5 to get rid of the fractions,

$$5(54) - 5(24) + \cancel{5}.\ \frac{36}{\cancel{5}} = \cancel{5}.\ \frac{216}{\cancel{5}} - 5.\ 6$$

$$270 - 120 + 36 = 216 - 30$$
$$186 = 186$$

Practice-5: Solve and Check

$$3(x - 8) - 12 = 13 - 2(x + 3)$$

Linear Equations with Fractions: The easiest way to solve linear equations with fractions is To get rid of the fractions by multiplying the whole equation by the least common denominator (LCD).

Example-6	Solve the equation: $\dfrac{x-7}{3} + \dfrac{2x+1}{6} = \dfrac{5x+3}{3}$

Solution $\quad \dfrac{x-7}{3} + \dfrac{2x+1}{6} = \dfrac{5x+3}{3}$

The LCD for 3, 6, and 3 is 6 →multiply the whole equation by 6,

$$\overset{2}{\cancel{6}} \cdot \dfrac{x-7}{3} + \cancel{6} \cdot \dfrac{2x+1}{\cancel{6}} = \overset{2}{\cancel{6}} \cdot \dfrac{5x+3}{\cancel{3}}$$

Cancel the denominators to get,
2(x–7) + 2x+1 = 2(5x+3)
Open parenthesis and use distribution to get,

2x – 2(7) + 2x + 1 = 2 (5x) + 2.3

2x – 14 + 2x +1 = 10x + 6

2x+2x –14 + 1 = 10x + 6

4x – 13 = 10x + 6

4x – 10x = 13 + 6

– 6x = 19, then → x = $\dfrac{-19}{6}$, or solution set is $\{ \dfrac{-19}{6} \}$={–3.2}

Graphical Solution: Using TI-83
Enter the left side on Y1 = (x–7)/3 + (2x+1)/6
Enter the right side on Y2 = (5x+3)/3

X	Y1	Y2
-3.3	-4.367	-4.5
-3.2	-4.3	-4.3
-3.1	-4.233	-4.167

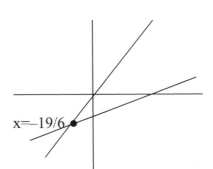

x=—19/6

Practice-6: Solve and Check

$$\frac{x+5}{3} + \frac{x-3}{6} = \frac{5x+3}{12}$$

Example-7

Solve the equation: $\dfrac{3x+1}{4} - 2 = \dfrac{x+1}{3}$

Solution: $\dfrac{3x+1}{4} - 2 = \dfrac{x+1}{3}$

The first step is to get rid of the denominator by multiplying the equation by the LCD = 12

$$\overset{3}{\cancel{12}}\cdot\frac{3x+1}{\cancel{4}} - 12\cdot 2 = \overset{4}{\cancel{12}}\cdot\frac{x+1}{\cancel{3}}\ ,\text{ then cancel out the denominators to get,}$$

$3(3x+1) - 24 = 4(x+1)$

Distribution gives, $9x + 3 - 24 = 4x + 4$

$$9x - 21 = 4x + 4$$

$9x - 4x = 21 + 4 \rightarrow 5x = 25$, then $\rightarrow x = 5$ or the solution set is $\{5\}$

To check, substitute the value of x back into the original equation,

$$\frac{3(5) + 1}{4} - 2 \overset{?}{=} \frac{5 + 1}{3}$$

$$\frac{16}{4} - 2 \overset{?}{=} \frac{6}{3}$$

$4 - 2 = 2 \rightarrow$ then 2=2. A true statement

Practice-7: Solve and Check

$$\frac{2x + 1}{5} + 2 = \frac{x}{3}$$

Example-8	
	Solve the equation: $\dfrac{3x}{2} - \dfrac{x}{3} = \dfrac{x}{6}$

Solution: $\dfrac{3x}{2} - \dfrac{x}{3} = \dfrac{x}{6}$

Multiply both sides of the equation by LCD = 6 to get rid of the fractions:

$\cancel{6} \cdot \dfrac{3 \quad 3x}{\cancel{2}} - \cancel{6} \cdot \dfrac{2 \quad x}{\cancel{3}} = \cancel{6} \cdot \dfrac{x}{\cancel{6}}$ Then cancel out the denominators,

3. $3x - 2x = x$
    $9x - 2x = x$
        $7x = x$
    $7x - x = 0$
        $6x = 0$, then → $x = 0$, or solution set is {0}

To check, substitute x=0 back into the equation:

$$\dfrac{3(0)}{2} - \dfrac{(0)}{3} = \dfrac{(0)}{6}$$

This gives 0 = 0 true statement.

Practice-8: Solve and Check

$$\dfrac{7x}{4} - \dfrac{x}{3} = \dfrac{x}{2}$$

Example-9

Solve the equation: $\dfrac{2x}{5} - \dfrac{1}{5} = 5$

Solution: $\dfrac{2x}{5} - \dfrac{1}{5} = 5$

The first step is to get rid of the denominator by multiplying the equation by the LCD = 5

$$\cancel{5}\cdot\frac{2x}{\cancel{5}} - \cancel{5}\cdot\frac{1}{\cancel{5}} = 5\cdot5$$

Then cancel out the denominator gives,

$$2x - 1 = 25$$

$$2x = 25 + 1$$

$$2x = 26 \rightarrow x = \frac{26}{2} = 13 \text{ or the solution set is } \{13\}$$

To check substitute x = 13 back into the original equation and show that the RS = LS

$$\frac{2(13)}{5} - \frac{1}{5} = 5$$

$$\frac{26}{5} - \frac{1}{5} = 5$$

Multiply by the LCD = 5, and cancel out the denominators:

$$\cancel{5}\cdot\frac{26}{\cancel{5}} - \cancel{5}\cdot\frac{1}{\cancel{5}} = 5\cdot5$$

$26 - 1 = 25 \rightarrow 25 = 25$ True statement then, $\rightarrow$ x= 13 is the solution

---

Practice-9: Solve and Check

$$\frac{2x}{3} - \frac{5}{3} = 3$$

Example-10	Solve the equation: $\dfrac{3x-2}{3} = \dfrac{2x-5}{5}$

Solution: Multiplying by the LCD = 15, then cancelling out the denominators:

$$\frac{3x-2}{3} = \frac{2x-5}{5}$$

$$\cancel{15} \cdot \frac{3x-2}{\cancel{3}} = \cancel{15} \cdot \frac{2x-5}{\cancel{5}}$$

$$5(3x-2) = 3(2x-5)$$

$$15x - 10 = 6x - 15$$

$$15x - 6x = -15 + 10$$

$$9x = -5 \rightarrow \text{ then } x = \frac{-5}{9} \text{ or the solution set is: } \left\{\frac{-5}{9}\right\}$$

To check substitute $x = \dfrac{-5}{9}$ back into the original equation and show that the RS = LS

The proof is left for the students.

Practice-10: Solve and Check

$$\frac{7x+2}{4} = \frac{x-2}{3}$$

Type of Equations	1. The equation that is true for all real numbers for which both sides are defined is called an *Identity*, for example: $x + 4 = 3 + x + 1$ The solution set to this equation is all the real numbers written as: { x/ x is a real number} or $\Re$ 2. The equation that is not an Identity, but is true for at least one real Number is called a *conditional* equation, for example: $x + 3 = 6$, The equation is true only if $x = 3$ 3. The equation that is not true for even one real number is called *Inconsistent* equation or *contradiction*, for example $2x = 2x + 10$ This equation has no solution, its solution set is written as: { }, or Ø

1.3 Application of Linear Equations

Application of Linear Equation in One Variable

Consecutive Integers Property	1. Consecutive integers: x, x+1, x+2 . . . 2. Consecutive odd integers: x, x+2, x+4, x+6 . . . 3. Consecutive even integers: x, x+2, x+4, x+6 . . .

Application of Integers:

Example-1	The sum of two consecutive integers is 31. What are the two integers?

Solution: Let the first integer be x, then

The second consecutive integer is x+1, and

The linear equation is → $x + (x+1) = 31$, which have the following solution:

$2x = 31 - 1 \rightarrow 2x = 30$.

Then the first integer is $\rightarrow x = 15$, and the second consecutive integer is $15+1 = 16$.
And the sum of the integers $= 15+16 = 31$.

Practice-1: Solve and Check
The sum of two consecutive integers is 51. What are the two integers?

Example-2	
	The sum of three odd consecutive integers is 57. What are the three integers?

Solution: Let the first integer $= x$, then

The second odd consecutive integer $= x+2$, and

The third consecutive integer $= x+4$. Then the linear equation is:

$$x + (x+2) + (x+4) = 57, \text{ with the}$$

The following solution:

$$x + x + 2 + x + 4 = 57$$
$$3x + 6 = 57$$
$$3x = 57 - 6 \rightarrow 3x = 51$$

Then the first integer is $x = 51/3 = 17$

The second odd consecutive integer $= 17 + 2 = 19$.

The third odd consecutive integer is $= 17 + 4 = 21$

The sum of all three odd consecutive integers is $= 17+19+21 = 57$

Practice-2: Solve and Check
The sum of three even consecutive integers is 60. What are the three Integers?

Application of Finance:

Interest Formula	I = Prt, where, p= Principal, r = % rate, t=time in years

Example-3	
	David wants to invest his $3000 by depositing part of it in bonds at 3%, and the rest in savings account at 2%, his interest after one year was $80, what amount did he invest in each account?

Using the above formula for interest: I = Prt
Let the amount deposited at 3% be = x, then → 3%x = the total interest at 3%.
The amount deposited in savings at 2% = 3000 – x, with interest = 2%(3000 – x)
Then his total interest on both accounts in one year = $80
The linear equation for the problem is: 3% x + 2%(3000–x) = 80

Solving the linear equation gives:

$0.03 x + 0.02 (3000) – 0.02 x = 80$

$0.03 x + 60 – 0.02 x = 80$

$0.01 x = 80 – 60$
$0.01x = 20$ → $x = 20/.01 = 2000$ the amount invested at 3%,
Then the rest of the amount that is invested at 2% = 3000-2000= $1000.

Practice-3	
	David wants to invest his $5000 by depositing part of it in bonds at 3%, and the rest in saving accounts at 2%, his interest after one year was $90, what amount did he invest in each account?

Example-4	
	Alana studied twice as many times as Tommy. Viviane studied 3hrs more than Tommy. If they all studied 31 hrs. How many hours did each study?

Solution: Let the time of study for Tommy = x,

Then Alana studied 2x, and Viviane studied (x+3)

The total hours they studied = 31, then the linear equation for the problem is:

$x + 2x + (x+3) = 31$ solving for x:

$$4x + 3 = 31$$
$$4x = 31 - 3$$
$$4x = 28 \rightarrow \text{then } x = 28/4 = 7 \text{ hrs}$$

Then Tommy studied 7 hours.

Alana's time of study = 2.7 = 14 hrs, and

Viviane's time = 7+3 = 10

And their totals = 31.

Practice-4	
	Mark studied for math finals 4 hrs less than Ronnie. David studied 2 hrs more than Mark. If they all studied total of 12 hrs. How many hours did each study?

Application of Geometry:

Geometry Formula's	Rectangle area = length x width Circle area = πr^2 Square area = length^2 Triangle area = ½ base x height Perimeter of rectangle p = 2(length + width) Circumference of circle = $2\pi r$ Total angles of a triangle sum up to 180°

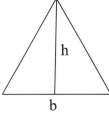

Area = 1/2 b h

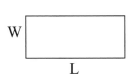

Area = LW

Area= πr^2

Example-5	If the length of a rectangle is 3times its width, and its perimeter is 45 cm Find its dimension.

Solution: Let the width of rectangle = x, then

The length of rectangle = 3x

The linear equation of the perimeter is:

P = 2L + 2W

45 = 2(3x) + 2x

Solving this linear equation for x:

45 = 6x + 2x

45 = 8x, → then x = 45/8 = 5.625 cm

Then the dimensions of the rectangle are:
Width = x = 5.625 cm
Length = 3x = 3(5.625) =16.875 cm, and the perimeter is = 2(5.625) + 2(16.875) = 45 cm.

Practice-5	
	If the length of a rectangle is 2 more than 3times its width, and its perimeter is 75 cm Find its dimension.

Application of Distance:

Distance Formula	Distance = Rate • Time

Example-6	
	Ronnie and Kelsey start running at the same time, if Ronnie is twice faster than Kelsey, what speed are they running to cover a distance of 30 miles in 2-hours?

Solution: Let the speed of Kelsey = x miles/hr, then
The speed of Ronnie = 2x
The linear equation of the distance is:
d = r t
30 = x (2) → x = 30/2 = 15 mi/hr Kelsey's speed.
Then Ronnie Runs with speed = 2x = 2(15) = 30 mi/h

Practice-6	
	Ronnie driving a car, leaves the city going north at a rate of 45 mi/hr at 2 pm. One hour later Kelsey drives going south of the city at rate of 35 mi/hr. At what time will they meet at 300 miles?

1.4 Exponents and Their Properties

On this section we will go over few examples, assuming that the students are familiar with this section from previous course in math.

Exponents Formula's	1. Multiplication rules: $x^n \cdot x^m = x^{n+m}$
	2. Quotient Rule: $\dfrac{x^n}{x^m} = x^{n-m}$
	3. Power: $(x^n)^m = x^{nm}$
	4. Inverse: $\dfrac{1}{x^n} = x^{-n}$

Example-1	
	Simplify the exponential expressions:
	a. $(-7 x^2 y^{-12})(-5 x^3 y^7)$
	b. $\left(\dfrac{20 x^3 y^6}{5 x^8 y^{-4}}\right)^3$
	c. $\left(\dfrac{x^{-3} y^7}{3}\right)^{-3}$

Solution: using the above rules we get the solution as follows:

a. $(-7 x^2 y^{-12})(-5 x^3 y^7) = \{(-7)(-5)\}\ \{(x^2)(x^3)\}\ \{(y^{-12})(y^7)\}$

$$= 35\ x^{2+3}\ y^{-12+7} = 35\ x^5\ y^{-5}$$

b. $\left(\dfrac{20\ x^3\ y^6}{5\ x^8\ y^{-4}}\right)^3 = ((20/5)\ x^{3-8}\ y^{6-(-4)})^3 = (4\ x^{-5}\ y^{10})^3 = 4^3\ x^{-15}\ y^{30}$

c. $\left(\dfrac{x^{-3}\ y^7}{3}\right)^{-3} = \dfrac{x^{(-3)(-3)}\ y^{(7)(-3)}}{3^{-3}} = \dfrac{x^9\ y^{-21}}{3^{-3}} = \dfrac{3^3\ x^9}{y^{21}} = \dfrac{27\ x^9}{y^{21}}$

Practice -1	Simplify the exponential expressions: a. $(-6 x^4 y^{-11})(-3 x^2 y^{-3})^2$ b. $\left(\dfrac{30\ x^5\ y^{-5}}{5\ x^9\ y^{-6}}\right)^2$ c. $\left(\dfrac{x^{-2}\ y^{-14}}{2}\right)^{-1}$

1.5 Scientific Notations

Objectives:	1. Converting from Scientific to Decimal Notations
	2. Converting from Decimal to Scientific Notations.

Rules	The scientific notation is expressed as:$(a10^n)$, where a is the number, n is the number of places the decimal point was moved, and is an integer: 1. $n > 0$ if decimal point was moved to the left. 2. $n < 0$ if the decimal point was moved to the right. 3. $n = 0$ if the decimal point was not moved. To convert to decimal, move the decimal point n-times to the right for $n > 0$

1. Converting to Decimal Notations

Example-1	Write each number in Decimal notation: a. 3.5×10^5 b. -7.3×10^4 c. 9.012×10^{-3}

Solution: a. $3.5 \times 10^5 \xrightarrow{\text{Move 5 to R}} 350000.$

b. $-7.3 \times 10^4 \xrightarrow{\text{Move 4 to R}} 73000.$

c. $9.012 \times 10^{-3} \xrightarrow{\text{Move 3 to L}} 0.009012$

Practice-1	Write each number in Decimal notation: a. 3.235×10^5 b. -73.21×10^4 c. 90.12×10^{-5}

2. Converting to Scientific Notations

Example-2	Write each number in scientific notation: a. 53,678,000,000,000,000 b. 0.000000000000723 c. $-$ 0.0000000000703

Solution: a. $53{,}678{,}000{,}000{,}000{,}000 \longrightarrow 5.3678 \times 10^{16}$

b. $0.000000000000723 \longrightarrow 7.23 \times 10^{-13}$

c. $-0.0000000000703 \longrightarrow -7.03 \times 10^{-11}$

Practice-2	Write each number in Scientific notation: a. 7,320,000,000,0 b.− 0.00000000965

These problems can be done using TI after changing the mode as shown:

Using TI: Change the Mode to → SCI
　　　　　 Then enter the decimal number
　　　　　 Then press enter

Chapter-1 Exercise

Solve and check the following linear equations:

1. $2x - 5 = 7$

2. $5x + 20 = 4x$

3. $2 - 3(7-x) = 5x + 7$

4. $2(x-4) + 3(x+7) = 2x-2$

5. $\dfrac{2x}{3} = \dfrac{x}{6} + 3$

6. $\dfrac{2x}{3} = -\dfrac{x}{4} + 6$

7. $\dfrac{4x+2}{4} - \dfrac{15}{3} = \dfrac{2-x}{2}$

8. $\dfrac{x}{2} - \dfrac{2}{10} = \dfrac{2x}{2} + \dfrac{2}{5}$

Solve each equation, and state whether the equation is an identity; a conditional; or an inconsistent equation

9. $7x + 5 = 5x + 15 + 2x$

10. $4x - 10 + 3x + 23 = 13 + 7x$

11. $5x - 10 + x = 5 + 4(x - 3)$

12. $4(x-2) - 3(x+5) = 2(2x-3) - 3(x+1)$

Solve the following word problems:

13. One angle of a triangle measures 20 more than the second angle. The measure of the third Angle is twice the sum of the measure of the first two angles. Determine the measure of each angle.

14. The weekly production cost C of manufacturing x watches is given by the formula
    $C(x) = 4500 + 2x$, where the variable C is in dollars:
    (a) What is the cost of producing 1000 watches?
    (b) How many watches can they produce for the cost of $8000?

15. The distance from earth to the moon is 4×10^8 meters, express this distance as a whole Number.

16. The wavelength of a visible light is about 5×10^{-7} meter. Express this wavelength as a Decimal

17. Find the area of a rectangle with length 6 feet and width 3 feet.

18. Find the volume and surface area of a box with length 6 feet, width 2 feet, and height 5 feet.

19. One number is 6 more than another. If the sum of the smaller and twice the larger number is 42. Find the two numbers.

20. Find two consecutive odd integers such that 3 times the first integer is 5 more than twice the Second.

Chapter-1 Test

Solve and check the following linear equations:

1. $5 - 12x = 8 - 7x - [6 \div 2(2 + 5^3) - 5x]$

2. $8 + \dfrac{x-2}{4} = \dfrac{x+3}{4}$

3. $3x + 4 = x$

4. $-3x - (6x - 5) = 40$

5. $(x - 2) + 9 = 2(x + 5)$

6. $7 - 12x = 8 - 7x - [6 \div 2(2 + 5^3) - 5x]$

7. $5 + \dfrac{x-2}{4} = \dfrac{x+3}{8}$

8. $5x + 9 = 2x + 9$

9. $2[4x - (2x - 6)] = 4(x-6)$

10. $3(x-4) + x = 2(6 + 2x)$

2. Solving System of Equations

2. Solving System of Equations

Objectives: 1. Solving System of Equations with Two Variables
2. Solving System of Equations with Three Variables
3. Applications

Method of solutions	• Algebraic solution • Graphical Solution • Matrix Solution

Type of solutions	• A unique solution (one point of intersection between the two lines) • Many Solutions (more than one point of intersection between two Lines), Same lines. • No solution (lines never intersect) parallel lines. • Solution set is a pair of points (x, y)

Intersecting lines
System has one
solution unique
solution

Parallel lines
System has
No Solution

Coincident lines
System has infinite
number of solution

2.1 Solving System of Equations with Two Variables

Example-1	Solve the system of equations $\quad y = 4 - 2x$ $\quad 6x - 3y = 18$

Algebraic Solution: The easiest way to solve this system is using substitution method
Substitute y= 4–2x into the second equation:
The second equation becomes: 6x – 3(4–2x) = 18 solve this linear equation for the variable x:
$$6x - 12 + 6x = 18$$
$$6x + 6x - 12 = 18$$
$$12x - 12 = 18$$
$$12x = 18 + 12$$
$$12x = 30 \rightarrow x = 30/12 = 2.5$$
Now we need to find y: substitute x=2.5 into y = 4 – 2x
$$y = 4 - 2(2.5)$$
$$y = -1$$
Then the solution set is (x, y) = (2.5, –1) this is the point of intersection between the two lines.

Graphical solution: Rewrite the two equations in the standard forms
$$y = -2x + 4 \text{ enter as Y1}$$
And, $\quad 6x - 3y = 18$
$$-3y = -6x + 18$$
$$y = 2x - 6 \text{ let it = Y2}$$
Graph Y1 and Y2 and zoom at the point of intersection which is the solution of the set.

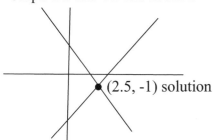
(2.5, -1) solution

Matrix solution: Write the system in standard matrix form x, y, and then constant:

$$2x + y = 4$$
$$6x - 3y = 18$$

$$\left|\begin{array}{cc|c} 2 & 1 & 4 \\ 6 & -3 & 18 \end{array}\right|$$ Using Ti-83: 2nd matrix → Edit enter 2 x 3(size)
2nd matrix → Math Scroll down to rref (matrix [A]) enter

The result shows as: $\left|\begin{array}{cc|c} 1 & 0 & 2.5 \\ 0 & 1 & -1 \end{array}\right|$ and is read as: x=2.5, y=−1

Practice-1	Solve the system of equations
	$y = 1 - 2x$
	$3x + y = 4$

Example-2	Solve the system of equations
	$1/2\ x - 3y = -1/2$
	$3x - 4y = 4$

Algebraic Solution: The easiest way to solve this system is first to simplify the first equation by multiplying it by 2 (the common denominator) to get rid of the fraction:

$$\cancel{2} \cdot \frac{1}{\cancel{2}}x - 2(3)y = \cancel{2} \cdot -\frac{1}{\cancel{2}}$$

Then the system is: $x - 6y = -1$
$$3x - 4y = 4$$

Using elimination method to solve, multiply the first equation by (-3) then add to the second equation: $-3x + 18y = 3$
$$3x - 4y = 4$$

$$14y = 7 \rightarrow \text{then } y = 7/14 = 1/2$$

Substitute y=1/2 into the first equation or the second one to get x = 2
The solution set is {x, y} = {1/2, 2}

Graphical solution: Rewrite the two equations in the standard forms
$$18y = 3x + 3$$
$$y = (1/6) x + 1/6 \text{ enter as Y1}$$
And, $3x - 4y = 4$
$$-4y = -3x + 4$$
$$y = 3/4 \ x - 1 \text{ let it = Y2}$$
Graph Y1 and Y2 and zoom at the point of intersection which is the solution of the set.
On TI-83 you'll find the solution as;   (2, 1/2)

X	Y1	Y2
1	.333	-.25
2	.5	.5
3	.66	1.25

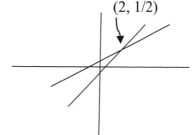

Matrix solution: Write the system in standard matrix form x, y, and then constant:
$$1/2x - 3y = -1/2$$
$$3x - 4y = 4$$

$\begin{vmatrix} 1/2 & -3 & -1/2 \\ 3 & -4 & 4 \end{vmatrix}$ Using TI-83: 2nd matrix → Edit enter 2 x 3(size)
2nd matrix → Math Scroll down to rref (matrix [A]) enter

The result shows as: $\begin{vmatrix} 1 & 0 & 2 \\ 0 & 1 & .5 \end{vmatrix}$ and is read as: x=2, y= 0.5

Practice-2	Solve the system of equations $2x - 5y = -3$ $3/2 \ x - 2 \ y = 5/2$

Example-3	Solve the system of equations
	$2x = 4 - 8y$ $3y = -2x + 16$

Algebraic Solution: Solve the first equation for x then substitute in the second equation

$x = 2 - 4y$

Then the second equation is: $3 = -2(2-4y) + 16$

$3 = -4 + 8y + 16$

Solving this equation for y gives $y = -12/5 = -2.4$

Substitute this back in $x = 2 - 4y$ gives $x = 11.6$

The solution set is $\{x, y\} = \{11.6, 2.4\}$

Graphical solution: Rewrite the two equations in the standard forms

$8y = -2x + 4$

$y = (-1/4) x + 1/2$ enter as Y1

And, $3y = -2x + 16$

$y = -2/3x + 16/3 = Y2$

Graph Y1 and Y2 and zoom at the point of intersection which is the solution of the set. On TI-83 you'll find the solution as;

X	Y1	Y2
11.5	-2.375	-2.33
11.6	-2.4	-2.4
11.7	-2.425	-2.4

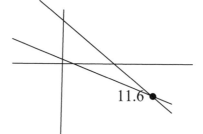

Matrix solution: Write the system in standard matrix form x, y, and then constant:

$2x + 3y = 4$

$2x + 3y = 16$

$$\begin{vmatrix} 2 & 8 & | & 4 \\ 2 & 3 & | & 16 \end{vmatrix}$$ Using TI-83: 2nd matrix → Edit enter 2 x 3(size)

2nd matrix → Math Scroll down to rref (matrix [A]) enter

The result shows as: $\begin{vmatrix} 1 & 0 & | & 11.6 \\ 0 & 1 & | & -2.4 \end{vmatrix}$ and is read as: x=11.6, y= –2.4

Practice-3	Solve the system of equations $2x - y = 5$ $5x + 2y = 8$

Example-4	Solve the system of equations $5x - 2y = 4$ $10x - 4y = 12$

Algebraic Solution: Multiply the first equation by (–2) then add to the second equation:

The system is: $-10x + 4y = -8$

$10x - 4y = 12$

$0 = 4$ false statement → System has no solution

The solution set is { } or Ø empty set

Graphical solution: Rewrite the two equations in the standard forms

$-2y = -5x + 4$

$y = (5/2)x - 2$ enter as Y1

And, $-4y = -10x + 12$

$y = 5/2x - 3 = Y2$

Graph Y1 and Y2 we get two parallel lines that will never intersect.

On TI-83 you'll find the solution as;

Matrix solution: Write the system in standard matrix form x, y, and then constant:

$$5x - 2y = 4$$
$$10x - 4y = 12$$

$$\left|\begin{array}{cc|c} 5 & -2 & 4 \\ 10 & -4 & 12 \end{array}\right|$$ Using TI-83: 2nd matrix → Edit enter 2 x 3(size)
2nd matrix → Math Scroll down to rref (matrix [A]) enter

The result shows as: $$\left|\begin{array}{cc|c} 1 & -.4 & 0 \\ 0 & 0 & 1 \end{array}\right|$$

The last raw of the matrix states that: 0 = 1 which is a false statement, this means
The system has no solution

Practice-4	Solve the system of equations $-y + 3x = 8$ $-2y + 6x = 10$

Example-5	Solve the system of equations $y = 1/2\ x - 4$ $2x - 4y = 16$

Algebraic Solution: Substitute the first equation in the second one:
The second equation becomes: $2x - 4(1/2\ x - 4) = 16$ simplifying gives:
$0 = 0$ True statement which means the system has infinite number of solutions, and the solution
set is {x/x all the real numbers} or$\Re$.

Graphical solution: Notice if you rewrite the second equation in the standard form you'll get
the first equation, it is the same line $y = 1/2\ x - 4$
Graph Y1 and Y2 we get the same line

Matrix solution: Write the system in standard matrix form x, y, and then constant:

$-1/2x + y = -4$
$2x - 4y = 16$

$$\begin{vmatrix} -1/2 & 1 & -4 \\ 2 & -4 & 16 \end{vmatrix}$$ Using TI-83: 2nd matrix → Edit enter 2 x 3(size)
2nd matrix → Math Scroll down to rref (matrix [A]) enter

The result shows as: $\begin{vmatrix} 1 & -2 & 8 \\ 0 & 0 & 0 \end{vmatrix}$

The last raw of the matrix states that: $0 = 0$ which is a true statement, this means
The system has infinite number of solutions

Practice-5	Solve the system of equations
	$y = 4x - 8$
	$5y - 20x = -40$

2.2 Solving System of Equations with Three Variables

Type of solutions	• A unique solution (one point of intersection between the three lines)
	• Many Solutions (more than one point of intersection between the Three lines), Same lines.
	• No solution (lines never intersect) parallel lines.
	• Solution set is a triplet point (x, y, z)
	Intersecting lines Parallel lines Coincident lines
	System has one System has System has infinite
	solution unique No Solution number of solution
	solution

Example-1	Solve the system of equations
	$x - y - z = 0$ ...... (1) $2y - 3z = -8$ ... (2) $3x + 3y + 2z = 4$ ....(3)

Algebraic Solution: We will use the elimination method as follows:
• Reduce the system of three equations with three variables into system of two equations
  With two variables
• Reduce the system of two equations with two variables into one equation with one variable.
• Use backward substitution to find the second variable.
• Substitute the two variables back into the first equation to find the third variable.

$$-3(1) \rightarrow -3x + 3y + 3z = 0$$
$$+(3) \rightarrow 3x + 3y + 2z = 4$$

$$6y + 5z = 4 .... (4)$$

$$-3(2) \rightarrow -6y + 9z = 24$$
$$+(4) \rightarrow 6y + 5z = 4$$

$$14z = 28 \rightarrow \text{then } z = 28/14 = 2$$

Now using backward substitution to get x and y.
$$(4) \rightarrow 6y + 5(2) = 4$$
$$6y + 10 = 4$$
$$6y = -6 \rightarrow \text{then } y = -1$$
Now substitute both y, and z into equation (1) or (3) to get x:
$$(1) \rightarrow x - y - z = 0$$
$$x + 1 - 2 = 0 \rightarrow \text{then } x=1 \text{ Solution set is } \{x, y, z\} = (1, -1, 2) \text{ a unique solution}$$

Matrix solution: Write the system in standard matrix form x, y, z and then constant:

$$x - y - z = 0$$
$$2y - 3z = -8$$
$$3x + 3y + 2z = 4$$

$$\begin{vmatrix} 1 & -1 & -1 & 0 \\ 0 & 2 & -3 & -8 \\ 3 & 3 & 2 & 4 \end{vmatrix} \longrightarrow \text{the result shows as:} \begin{vmatrix} 1 & 0 & 0 & 1 \\ 0 & 1 & 0 & -1 \\ 0 & 0 & 1 & 2 \end{vmatrix}$$

The solution is on the last column: x=1, y=-1, z=2

Practice-1	Solve the system of equations $3x - y - 3z = 22 \ldots\ldots (1)$ $x - 2y + z = 0 \ldots\ldots (2)$ $2x + 3y - z = 16 \ldots\ldots(3)$

Example-2	Solve the system of equations $-5x - 4y + z = -11 \ldots\ldots (1)$ $x - y + z = -2 \ldots\ldots (2)$ $x + 2y - z = 5 \ldots\ldots (3)$

Eliminating x variable:

$$(1) \rightarrow -5x - 4y + z = -11$$
$$+5(2) \rightarrow 5x - 5y + 5z = -10$$
$$\overline{\qquad\qquad\qquad\qquad\qquad}$$
$$-9y + 6z = -21 \ldots (4)$$

$$(2) \rightarrow x - y + z = -2$$
$$-(3) \rightarrow x - 2y + z = -5$$
$$\overline{\qquad\qquad\qquad\qquad}$$
$$-3y + 2z = -7 \ldots (5)$$

Eliminate y from (4) and (5):

$(4) \rightarrow -9y + 6z = -21$

$-3(5) \rightarrow 9y - 6z = 21$

———————————————

$0 = 0$, this means the system can have infinite number of solutions.

Matrix solution: Write the system in standard matrix form x, y, z and then constant:

$-5x - 4y + z = -11$

$x - y + z = -2$

$x + 2y - z = 5$

$$\begin{vmatrix} -5 & -4 & 1 & -11 \\ 1 & -1 & 1 & -2 \\ 1 & 2 & -1 & 5 \end{vmatrix} \longrightarrow \text{The result shows as: } \begin{vmatrix} 1 & 0 & .33 & .33 \\ 0 & 1 & -.66 & 2.3 \\ 0 & 0 & 0 & 0 \end{vmatrix}$$

The last raw states that 0=0 and this is a true statement which states that the system has infinite number of solutions.

Practice-2	Solve the system of equations
	$x - 2y - z = 8 ...... (1)$
	$4x - 5y + 5z = 53 .... (2)$
	$2x - 3y + z = 23 .... (3)$

Example-3	Solve the system of equations
	$2x - 3y - z = 0 ....(1)$
	$3x - 4y - z = 1 ....(2)$
	$-x + 2y + z = 5 ....(3)$

Algebraic Solution: We will use the elimination method as follows:

$-3(1) \rightarrow -6x + 9y + 3z = 0$
$+2(2) \rightarrow 6x - 8y - 2z = 2$

$\overline{\hspace{6cm}}$

$\qquad\qquad y + z = 2 \dots (4)$
$(1) \rightarrow 2x - 3y - z = 0$
$+2(3) \rightarrow -2x + 4y + 2z = 10$

$\overline{\hspace{6cm}}$

$\qquad\qquad y + z = 10 \dots (5)$

Now solving (4), and (5) gives:
$-(4) \rightarrow -y - z = -2$
$\qquad\qquad y + z = 10$

$\overline{\hspace{5cm}}$

$\qquad\quad 0 = 8$ this is a false statement, which means the system has no solution.
Or the solution set is { } or Ø for empty set.

Matrix solution: Write the system in standard matrix form x, y, z and then constant:

$2x - 3y - z = 0$
$3x - 4y - z = 1$
$-x + 2y + z = 5$

$$\left|\begin{array}{ccc|c} 2 & -3 & -1 & 0 \\ 3 & -4 & -1 & 1 \\ -1 & 2 & 1 & 5 \end{array}\right| \longrightarrow \text{the result shows as:} \left|\begin{array}{ccc|c} 1 & 0 & 1 & 0 \\ 0 & 1 & 1 & 0 \\ 0 & 0 & 0 & 1 \end{array}\right|$$

The last raw states that $0 = 1$ which is a false statement
This means the system has no solution, because the lines are parallel to each other.

Practice-3	Solve the system of equations
	$3x + 2y \quad\quad = 0 \ \dots (1)$
	$x - \ y - z = 1 \ \dots (2)$
	$2x + 3y + \ z = 2 \ \dots (3)$

2.3 Applications

Example-1	A chemist needs to mix 14% acid solution with 22% acid solution to produce 180 ounces of 17% acid solution. How many ounces of each acid solution must he uses?

Solution: Let x= the number of ounces of the first acid solution.

y= the number of ounces of the second acid solution

Solutions	%	Total amount
x	14%	0.14 x
y	22%	0.22 y
180	17%	0.17 (180)

The system of equations can be formed from:

The first column → x + y =180… (1), and

The third column → 0.14 x + 0.22 y = 30.6… (2)

Now we can solve the above system of two equations with two variables, using the steps before

Algebraic Solution: Solve equation (1) for y = 180 – x substitute into equation (2):

$$(2) \rightarrow 0.14 x + 0.22 (180 – x) = 30.6$$
$$0.14 x + 39.6 – 0.22 x = 30.6 \rightarrow x = 112.5 \text{ ounces}$$

Substituting back in (1) gives y=67.5 ounces.

Matrix solution: Write the system in standard matrix form x, y and then constant:

$$\left| \begin{array}{cc} 1 & 1 \\ 0.14 & 0.22 \end{array} \right| \begin{array}{c} 180 \\ 30.6 \end{array} \longrightarrow \left| \begin{array}{cc} 1 & 0 \\ 0 & 1 \end{array} \right| \begin{array}{c} 112.5 \\ 67.5 \end{array} \longrightarrow \begin{array}{l} x=112.5 \text{ ounces} \\ y=67.5 \text{ ounces} \end{array}$$

Practice-1	How much of 10% solution can you mix with 18% solution to form a mixture of 140 ounces of 13% solution?

Example-2	Alana wants to invest her $20,000 into 3-acounts, a treasure bills with 5% simple interest, treasure bonds with 7% simple interest, and a corporate bonds with 10% simple interest. Her wishes is to earn $1390/year in income, also wants her investment in treasure bills to be $3000 more that her investment in corporate bonds. How much money should Alana place in each account?

Solution: let the amount placed in the 3-accounts be x, y, and z as shown in the table.

Account	%	Total investment
x	5%	0.05 x
y	7%	0.07 y
z	10%	0.10 z
20,000		1390

The first two equations can be formed from the first column and the third column:

$x + y + z = 20,000$ … (1)

$0,05x + 0.7 y + 0.10z = 1390$ … (2)

And the third equation is: $x = z + 3000$ or → $x - z = 3000$ … (3)

Solving the three equations using Matrix:

$$\begin{vmatrix} 1 & 1 & 1 & 20,000 \\ 0.05 & 0.07 & 0.10 & 1390 \\ 1 & 0 & -1 & 3000 \end{vmatrix} \longrightarrow \begin{vmatrix} 1 & 0 & 0 & 8000 \\ 0 & 1 & 0 & 7000 \\ 0 & 0 & 1 & 5000 \end{vmatrix} \longrightarrow \begin{matrix} x=8000 \\ y=7000 \\ z=5000 \end{matrix}$$

Practice-2	Kathy wants to invest her $25,000 into 3-acounts, a treasure bills with 3% simple interest, treasure bonds with 5% simple interest, and a corporate bonds with 7% simple interest. Her wishes is to make a profit of $1200 /year, also wants her investment in treasure bonds to be $2000 more than the treasure bills. How much money should Kathy place in each account?

Example-3	Flying with the winds in a small airplane takes 480 miles in 2.5 hours. Flying against the wind, the same airplane might take 4 hours. Find the rate of the wind and the plane.

Solution: Using the distance formula $d = rt$ where r=rate, and t=time in, and assuming that:
x= rate of plane
y= rate of the wind

Flying	Rate	Time	Distance
With the wind	x+ y	2.5	2.5(x+ y) = 480
Against the wind	x– y	4	4(x– y) = 480

The system of equations is:
$2.5 x + 2.5 y = 480\ldots (1)$
$\ \ 4 x - \ \ 4 y = 480\ldots (2)$
Solving the system of two equations yields: x=156 mile/hour the rate of the plane
$\qquad\qquad\qquad\qquad\qquad$ y= 35 mile/hour the winds rate.

Practice - 3	Flying with the winds in a small airplane takes 580 miles in 4 hours. Flying against the wind, the same airplane might take 5 hours. Find the rate of the wind and the plane.

Chapter-2 Exercise

Solve and check the following system of two equations:

1. $x - y = 8$
    $x + y = 10$

2. $2 x + y = 4$
    $x - y = - 1$

3. $3x - 5y = 20$
    $4 x - 2y = 22$

4. $-2x + 4y = -4$
   $\quad x - \ y = 1$

Solve the following system of 3-equations:

5. $4x - y + 5z = 5$
   $\ 2x - y + 3z = 2$
   $\quad x - 2y + 3z = 1$

6. $\ x - 2y \quad = -7$
   $\ x + \ y + z = 6$
   $4x + 3y + z = 7$

7. $\ x + y \qquad = 1$
   $\qquad y + 2z = -2$
   $2x \qquad - z = 0$

8. $x + y \ + z = 8$
   $x \qquad - z = -4$
   $\qquad y + 3z = 12$

9. $x + y + \ z = 2$
   $\qquad - y + 2z = 1$
   $-x \qquad + 3z = 0$

10. $x - \ y \ + z = 0$
    $\qquad - 2y + z = -1$
    $-2x - 3y \qquad = -5$

For the following word problems, write the system of equations first then show the solution methods:

11. A chemist needs to mix 10% acid solution with 20% acid solution to produce 170 ounces of 16% acid solution. How many ounces of each acid solution must he uses?

12. Alana wants to invest her $10,000 into 3-acounts, a treasure bills with 5% simple interest, Treasure bonds with 7% simple interest, and corporate bonds with 10% simple interest Her wishes is to earn $695/year in income, also wants her investment in treasure bills to Be $1500 more that her investment in corporate bonds. How much money should Alana Place in each account?

13. Flying with the winds in a small airplane takes 400 miles in 1.5 hours. Flying against the wind, the same airplane might take 3 hours. Find the rate of the wind and the plane.

Chapter-2 Test

Solve and check the following system of two equations, and graph:

1. $2x - 5y = 6$
   $4x - 10y = 24$

2. $2x + 3y = 10$
   $y = 5 - x$

3. $x + y = 4$
   $x - y = 2$

4. $-2x + 3y = 5$
   $-2x + 3y = 5$

Solve and check the following system of three equations, and graph:

5. $x - y - z = 1$
   $2x + y - z = 4$
   $3x - y + z = 11$

6. $4x - y + 2z = 11$
   $x + 2y - z = -1$
   $2x + 2y - 3z = -1$

7. $3x - 2y + z = -7$
   $x + y - 2z = 12$
   $3x + y - z = 10$

8. $x + y + z = -1$
   $2x - y - 5z = 12$
   $-x + 2y + z = -3$

Solve the following linear inequalities:

9. $4(x+1) + 2 \geq 3x + 6$

10. $\dfrac{x}{10} - \dfrac{1}{5} \leq \dfrac{3x}{10} + 1$

3. Inequalities

3. Inequalities

Objectives: 1. Solving Linear Inequality
2. Solving Compound Inequality
3. Absolute Value Problems and Inequalities
4. Two Variable Inequality
5. Applications

3.1 Solving Linear Inequality

Solving linear inequalities is similar to solving linear equations except that using inequality symbols instead of equal signs, the table below shows these symbols:

Inequality Symbols and Their Rules	Solving Inequality problems involves the following symbols: < less the graph is represented by broken line > larger and the graph is represented by a broken line too ≤ less and equal represented by solid line ≥ Larger and equal the graph is Represented by a solid line too.

Here are some examples represented graphically with different symbols:

This table shows some other representation of symbols

Set	Interval	Graph
$x < c$	(∞, c)	←——————)———→ c
$x > c$	(∞, c)	←————(——————→ c
$x \geq c$	(∞, c)	←————[——————→ c
$x \leq c$	(∞, c)	←————]——————→ c
$a < x < b$	(a, b)	←——(———)——→ a b
$a < x < b$	(a, b)	←——(———]——→ a b
$a < x < b$	(a, b)	←——[———)——→ a b
$a < x < b$	(a, b)	←——[———]——→ a b

Example-1	Solve the Inequality: $\dfrac{2x}{3} < 1$

Solution: Multiply both sides of the inequality by 3

$$\cancel{3}.\ \dfrac{2x}{\cancel{3}} < 1\,(3)\ \rightarrow\ 2x < 3 \text{ divide both sides by 2 to get: } x < 3/2$$

Solution set graphically is: ←—————————)——→
                3/2

And the solution set is: $\{x/\ x < 3/2\}$ or $\{(-\infty, 3/2)\}$

Practice-1	Solve the Inequality: $\dfrac{3x}{5} < 2$

Example-2	Solve the Inequality: $2x \geq 5x + 18$

Solution: $2x \geq 5x + 18$

$\quad\quad\quad \underline{-5x \quad -5x}$

$\quad\quad\quad \dfrac{-3x}{-3} \geq \dfrac{18}{-3}$ Divide both sides by (-3) (Don't forget to switch the symbol)

$\quad\quad x \leq -6 \rightarrow$

$\quad\quad\quad\quad\quad\quad -6$

And the solution set is: $\{x/ \ x \leq -6\}$ or solution set is: $\{ (-\infty, -6)\}$

Practice-2	Solve the Inequality: $3x \geq 7x - 28$

Example-3	Solve the Inequality: $4(x +1) \geq 3x + 6$

Solution: $4(x +1) \geq 3x + 6$

$\quad\quad\quad 4x + 4 \geq 3x + 6$

$\quad\quad\quad 4x - 3x \geq 6 - 4$

$\quad\quad\quad x \geq 2$ the solution set is:

$\quad\quad\quad\quad\quad\quad\quad\quad\quad 2$ $\{[2, \infty)\}$

Practice-3	
	Solve the Inequality: $8x + 3 \geq 3(2x + 1)$

Example-4	
	Solve the Inequality: $6(x +4) – 12 < 11 + 12(3 + x)$

Solution: $6(x +4) – 12 < 11 + 12 (3 + x)$
$\qquad 6x + 24 – 12 < 11 + 36 + 12x$
$\qquad 6x + 12 < 47 + 12x$
$\qquad 6x – 12x < 47 – 12$
$\qquad –6x < 35$ → then $x > – 35/6$ → solution set is: $\{(–35/6, \infty)\}$

Practice-4	
	Solve the Inequality: $7(x +4) – 13 < 12 + 13(3 + x)$

Example-5	
	Solve the Inequality: $\dfrac{7}{2}(6 – 20x) – 3 > 5 – \dfrac{11}{2}(x–1)$

Solution: Multiply both sides of the inequality by 2 to get
$\qquad 7(6 – 20x) – 6 > 10 – 11(x– 1)$
$\qquad 42 – 140x – 6 > 10 – 11x + 11$
$\qquad 36 – 140x > 21 – 11x$
$\qquad –140x + 11x > 21 – 36$
$\qquad –129x > – 15$ → solution is $x< 15/129$ → $\{(–\infty, 15/129)\}$

Practice-5	
	Solve the Inequality: $\dfrac{5}{3}(6 + 10x) – 3 > 5 – \dfrac{12}{3}(x–2)$

3.2 Solving Compound Inequality

Compound Inequality	Two inequalities connected by a connector, is called compound The connectors are AND with symbol ∩ OR with symbol ∪

Definition	The symbol AND ∩ represents intersection between two sets. The symbol OR ∪ represents union between two sets Then A ∩ B = { x/ x∈A AND x ∈ B } And A ∪ B = { x/ x∈A OR x∈ B or in both}

Example-1	Find the intersection between the two sets: A = { 1, 2, 3, 4, 5, 6, 7} B = { 3, 4, 7, 8}

Solution: the intersection between A and B is:

$$A \cap B = \{1, 2, 3, 4, 5, 6, 7\} \cap \{3, 4, 7, 8\}$$
$$= \{3, 4, 7\}$$

Practice-1	Find the intersection between the two sets: A = { 3, 5, 7, 9, 10, 12, 17} B = { 3, 4, 7, 17}

Example-2	Find the union between the two sets: A = { 1, 2, 3, 4} B = { 3, 4, 7, 8}

Solution: the union between A and B are:

$$A \cup B = \{1, 2, 3, 4, 5, 6, 7\} \cup \{3, 4, 7, 8\}$$
$$= \{1, 2, 3, 4, 7, 8\}$$

Practice-2	Find the union between the two sets: A = { 3, 5, 7, 9} B = { 3, 4, 7, 17}

Union ($\cup$) and Intersection ($\cap$) of inequalities

Example-3	Find the intersection between the two sets: $2x - 1 < 5$ AND $x + 5 < 7$

Solution: Solve each inequality separately

$$2x - 1 < 5 \quad \cap \quad x+5 < 7$$
$$\underline{+1 \quad +1 \qquad -5 \quad -5}$$

$$\frac{2x}{2} < \frac{6}{2} \quad \cap \quad x < 2$$

$$x < 3 \cap x < 2$$

Then the solution can be represented graphically as follows:

$x < 3$
    3

$x<2$
   2

The solution of $x< 3 \cap x< 2$ is or $\{(-\infty, 2)\}$
                                  2

Practice-3	Find the intersection between the two sets: $1 - 3x < 6 - 5x$ AND $x-9 < 4x -3$

Example-4	Find the union between the two sets: $2x + 3 \leq 5$ OR $3x-7 \geq 8$

Solution: As before we will solve each inequality separately:

$$2x + 3 \le 5 \ \cup \ 3x - 7 \ge 8$$
$$-3 \quad -3 \qquad +7 \ +7$$

_____

$$2x \le 2 \quad \cup \quad 3x \ge 15$$

The solution of $x \le 1 \cup x \ge 5$ can be found as follows:

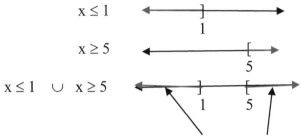

Then the solution set is $\{x \le 1 \cup x \ge 5\}$ or $\{(-\infty, 1) \cup (5, \infty)\}$

Practice-4	Find the union between the two sets: $2x + 1 < 15$ OR $3x - 4 \ge -1$

Definition	If $a < b$, then the compound inequality $\quad a < x$ AND $x < b$ can be written as: $\quad\quad a < x < b$

Example-5	Solve the inequality: $2 < 2x - 2 < 18$

Solution: $2 < 2x - 2 < 18$

$\phantom{2 <}+2 \quad +2 \quad +2$

$\rule{4cm}{0.4pt}$

$\phantom{2 <}4 < 2x < 20$

$\phantom{2 <}\dfrac{}{2} \quad \dfrac{}{2} \quad \dfrac{}{2}$

$2 < x < 10 \rightarrow$ solution set also $\{(2, 10)\}$

Practice-5	Solve the inequality: $3 \leq x/3 - 3 < -2$

Example-6	Solve the inequality: $-3 < \dfrac{2x - 3}{2} < 3$

Solution:

$$-3 < \dfrac{2x - 3}{2} < 3$$

Multiply all sides by 2:

$-6 < 2x - 3 < 6$ add 3 to all sites

$\phantom{-6 <}+3 \quad +3 \quad +3$

$\rule{4cm}{0.4pt}$

$-3 < 2x < 9$ divide by 2 gives

$-3/2 < x < 9/2 \rightarrow$ Solution set also $\{(-3/2, 9/2)\}$

Practice-6	Solve the inequality: $$-1 < \frac{3x - 5}{3} < 3$$

3.3 Absolute value Problems and Inequality

Absolute Value Rules	1. If $|x| = c \rightarrow x = \pm c$ 2. If $|x| < c \rightarrow -c < x < c$ 3. If $|x| > c \rightarrow x > c$ or $x < -c$ 4. If $|x| = |y| \rightarrow x = \pm y$ 5. $|x| < -c$ has no solution, or the solution set is { } or $\emptyset$ empty set. 6. $|x| > -c$ the solution set is all the real numbers $(-\infty, \infty)$

Example-1	Solve the absolute Value problem: $|3x - 2| = 7$

Solution: Applying rule (1) gives:
$$3x - 2 = 7 \text{ or } 3x - 2 = -7$$
Solving the two equations separately gives:
$$3x = 9 \quad \text{or} \quad 3x = -5$$
Or x = 9/3, and x = – 5/3

Practice -1	Solve the absolute Value problem: $|3x + 4| = 10$

Example-2	Solve the absolute Value Inequality problem: $|5x - 2| > 13$

Solution: Applying rule (3) gives:

$$5x - 2 > 13 \text{ or } 5x - 2 < -13$$

Solving the two inequality separately gives:

$$5x > 15 \quad \text{or} \quad 5x < -11$$

Then $x > 3 \cup x < -11/5$

Practice -2	Solve the absolute Value Inequality problem: $\quad \mid 3 - 5x/4 \mid > 9$

3.4 Two Variable Inequality

Two Variable Inequality	The two variable inequality is solved graphically because the solution set is a whole region. Steps of solving the 2-variable inequality: 1. Rewrite the inequality in the standard form $y < mx + b$ 2. Change the symbol into equal sign. 3. Graph the linear equation 4. shade the required region 5. Show the solution.

Example-1	Solve the linear inequality: $\quad 3x + y \leq 6$

Solution: Rewrite in the standard form: $y \leq -3x + 6$
Rewrite the equality: $y = -3x + 6$
Graph the line, and show the solution,

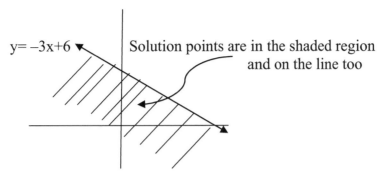

y= –3x+6

Solution points are in the shaded region and on the line too

Practice -1	Solve the linear inequality: x + 2y ≤ 2

Graphing a system of linear Inequalities

Example-2	Solve the system of linear inequalities: 2x + y ≤ 4..(1) x + y ≤ 3... (2)

Solution: (1) → y ≤ –2x + 4 → y = –2x + 4

(2) → y ≤ – x + 3

Now we graph and shade the required region:

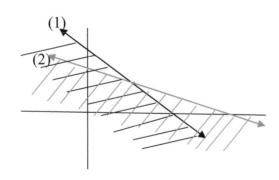

Practice -2	Solve the system of linear inequalities: $4x - y \geq 2 \ldots (1)$ $x + y \geq 3 \ldots (2)$

3.5 Applications

Linear programming is the best application for the system of linear inequalities, it is used in business, and faineance planning called optimization which represents minimization, and maximization. Each problem contains two parts:
• The objective equation written as: $z = ax + by$, where a, and b are real numbers.
• The constraints, which is a system of inequalities.

Example-1	Minimize the objective: $z = 2x + 3y$ Subject to the following constraints: $y \leq 5 \ (1)$ $x \leq 6 \ (2)$ $x + y \geq 2 \ (3)$ $x \geq 0 \ (4)$ $y \geq 0 \ (5)$

Solution: 1. Rewrite all the inequalities in the standard form and write their equations:

$$y \leq 5 \ \rightarrow \ y = 5$$
$$x \leq 6 \ \rightarrow \ x = 6$$
$$y \geq -x + 2 \ \rightarrow \ y = -x + 2$$
$$x \geq 0 \ \rightarrow \ x = 0$$
$$y \geq 0 \ \rightarrow \ y = 0$$

2. Graph and shade the required regions:

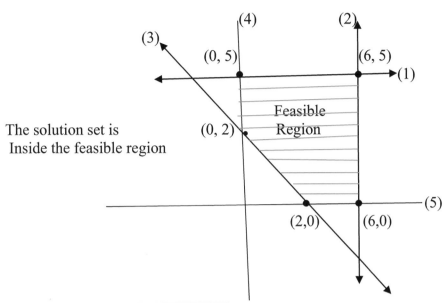

The solution set is
Inside the feasible region

Corner points	z = 2x + 3 y
(2,0)	z = 4
(6,0)	z=12
(6,5)	z=27
(0,5)	z=15
(0,2)	z=6

From the table we can see the minimum occurs at the point (2, 0)

Practice -1	Minimize the objective: z = 3x + 4y Subject to the following constraints: $x + y \le 8$ (1) $3x + 2y \ge 6$ (2) $x \ge 0$ (3) $y \ge 0$ (4)

Chapter-3 Exercise

Solve and graph the following inequalities:

1. $\dfrac{7x}{3} < 7$

2. $-3x + 2y \leq 6$

3. $1/2\, x - 3/2\, y < -10$

4. $x - y \geq 3$

5. $15 - 3x \leq 12 + 4y$

6. $3(2x-2) + 4 \leq -2(y+1)$

Solve and graph the following system of inequalities:

7. $2x+y \geq 2$
   $x-y \geq 5$

8. $-3x - 2y \leq 7$
   $-x - y \geq 9$

9. Graph the following inequalities:

   a. $x > 3$

   b. $x < -1$

   c. $y \geq 3$

   c. $y \leq 10$

10. Maximize the objective $z = x+2y$

   Subject to the following constraints:

   $x \geq 0,\ y \geq 0,\ x + y \leq 3,\ 2x - 4y \geq 6$

Chapter-3 Test

Solve and graph the inequalities:

1. $3 \left(\dfrac{3x - 12}{5} \right) - 13 \geq 2 \left(\dfrac{7y + 11}{3} \right) + 7$

2. Graph and give the solution of the system

$$14x + 7y \geq 21$$
$$x - 12y \leq 6$$

3. Solve the compound inequality: $-4 < 3x + 6 < 4$

4. Solve and graph: $-4x + 4y > 11$

5. Minimize the objective $z = x + 2y$ subject to the constraints:

$$x \geq 0, \ y \geq 0, \ x + y \geq 3, \ y \leq 3, \ x \leq 4$$

Graph the solution set of the system of inequalities or indicate that the system has no solution:

6. $-x + 2y \leq -6$
   $3x + 2y > -18$

7. $x \geq 0$
   $y > 0$
   $x + y \leq 7$
   $x + y \geq 6$

Solve the following inequalities

8. $(x+2)(x-3) > (x-1)(x+3)$

9. $x(9x - 5) \leq (3x - 1)^2$

10. For the given objective: $Z = 2x + 4y$ with the following constraints:

$$x \geq 0$$
$$y \geq 0$$
$$x + 3y \geq 6$$
$$x + y \geq 3$$
$$x + y \leq 9$$

a) Graph the system of inequalities and show the region of solution.
b) Find the values of the objective at the corners of the region of solution.
c) Determine the maximum value.

4. Polynomials

4. Polynomials

Objectives: 1. Introduction to Polynomials
2. Factoring Polynomials
3. Solving Polynomial Equations
4. Division of Polynomials
5. Solving Polynomial Inequalities
6. Applications

4.1 Introduction to Polynomials

Definition of Polynomials Functions	Polynomial functions can be described in the standard form as: $f(x) = a_n x^n + a_{n-1} x^{n-1} + \ldots + ax_1 + a_0$ Where: n is the degree of the polynomial a_n is the leading coefficient of the polynomial a_0 is the constant of the polynomial
Rules of Polynomials	Multiplication $\rightarrow$ $x^n . x^m = x^{n+m}$ Division $\rightarrow$ $x^n / x^m = x^{n-m}$ Power $\rightarrow$ $(x^n)^m = x^{nm}$ $x^{-n} = 1/x^n$ $1/x^{-n} = x^n$
Parts of Polynomials	One term of the polynomial is called Monomial Two terms of the polynomial is called Monomial Three terms of the polynomial is called Trinomial..etc

4.2 Factoring Polynomials

Objectives: • Factor the difference of two squares
  • Factor the difference and sum of two cubes
  • Factor perfect squares
  • Factor a second degree polynomial
  • Factor by grouping

Difference of two squares	$x^2 - y^2 = (x-y)(x+y)$
Sum of two cubes	$x^2 + y^2 = (x+y)(x^2 - xy + y^2)$
Difference of two cubes	$x^2 - y^2 = (x-y)(x^2 + xy + y^2)$
Perfect squares	$x^2 + 2xy + y^2 = (x+y)^2$ $x^2 - 2xy + y^2 = (x-y)^2$

Example -1	Factor the following: a. $x^2 - 9$ b. $x^3 + 27$ c. $x^3 - 27$ d. $a^2 + 2ab + b^2$ e. $a^2 - 2ab + b^2$

Solution: a. this binomial is a difference of two squares, and can be factored
  as follows: $x^2 - 9 = (x^2 - 3^2) = (x-3)(x+3)$
  b. is the sum of two cubes: $x^3 + 27 = (x^3 + 3^3) = (x+3)(x^2 - 3x + 3^2)$
  c. is the difference of two cubes: $x^3 - 27 = (x^3 - 3^3) = (x-3)(x^2 + 3x + 3^2)$

d. is a perfect square: $a^2 + 2ab + b^2 = (a + b)^2$
e. is a perfect square: $a^2 - 2ab + b^2 = (a - b)^2$

Practice -1	Factor the following: a. $x^2 - 81$ b. $x^3 + 8$ c. $x^3 - 8$ d. $9a^2 + 12ab + 4b^2$ e. $9a^2 - 12ab + 4b^2$

Example -2	Factor the second degree polynomials: a. $x^2 + 7x + 10$ b. $x^2 - 8x - 9$ c. $x^2 - 10x + 21$ d. $x^2 - 3x - 54$

Solution: a. $(x+2)\,(x+5)$
       b. $(x+1)\,(x-9)$
       c. $(x-7)\,(x-3)$
       d. $(x+6)\,(x-9)$

Practice -2	Factor the second degree polynomials: a. $3x^2 + 4x + 1$ b. $3x^2 - 2x - 8$ c. $3x^2 - 10x + 8$ d. $x^2 - 17x + 16$

Example -3	Factor by grouping: a. $6x^2 + 9x + 4x + 6$ b. $3x^2 - 6x + x - 2$ c. $2x^2 - 2x + 3x - 3$

Solution: a. $6x^2 + 9x + 4x + 6 = 3x(2x + 3) + 2(2x + 3)$
$$= (2x+3)(3x+2)$$

b. $3x^2 - 6x + x - 2 = 3x(x - 2) + (x-2)$
$$= (x-2)(3x +1)$$

c. $2x^2 - 2x + 3x - 3 = 2x(x-1) + 3(x-1)$
$$= (x-1)(2x +3)$$

Practice -3	Factor by grouping: a. $9x^2 + 9x + 4x + 4$ b. $2x^2 - 4x + x - 2$ c. $x^2 - x + 3x - 3$

4.3 Solving Polynomial Equations

In this section we will concentrate on polynomials of second degree (n=2) and their solutions by Factoring and grouping, these polynomials are also called the quadratic equations, which has the following standard form:

$$ax^2 + bx + c = 0$$

Solving Polynomial equations by factoring and Grouping

Example -1	Solve the quadratic equation by factoring: $4x^2 - 256 = 0$

Solution: $4x^2 - 256 = 0$
$$4(x^2 - 64) = 0$$
$$\text{Or} \rightarrow x^2 - 64 = 0$$

This is a difference of two squares written as: $x^2 - 8^2 = 0$

Which can be solved as: $(x-8)(x+8) = 0$ → solution set is $\{-8, 8\}$

Practice -1	Solve the quadratic equation by factoring: $2x^2 - 8 = 0$

Example -2	Solve the quadratic equation by factoring: $2x^2 - 6x - 20 = 0$

Solution: $2x^2 - 6x - 20 = 0$

$2(x^2 - 3x - 10) = 0$

Or → $x^2 - 3x - 10 = 0$ Factoring we get: $(x-5)(x+2) = 0$

And → solution set is $\{-2, 5\}$

Practice -2	Solve the quadratic equation by factoring: $2x^2 + 8x - 64 = 0$

Example -3	Solve the quadratic equation: $x(4x - 6) = 18$

Solution: $x(4x - 6) = 18$

$4x^2 - 6x - 18 = 0$

$2(2x^2 - 3x - 9) = 0$

Or $2x^2 - 3x - 9 = 0$ Factoring we get: $(2x+3)(x-3) = 0$

And → solution set is $\{-3/2, 3\}$

Practice -3	Solve the quadratic equation: $3x(x+2) = 45$

Example -4	Solve the quadratic equation: $(x - 1)(x - 3) = 15$

Solution: $(x-1)(x-3) = 15$

$$x^2 - 4x + 3 = 15$$
$$x^2 - 4x + 3 - 15 = 0$$
$$x^2 - 4x - 12 = 0$$

Factoring we get: $(x+2)(x-6) = 0$. And → solution set is $\{-2, 6\}$

Practice -4	Solve the quadratic equation: $(x - 2)(x + 3) = 14$

Example -5	Solve the quadratic equation by grouping: $x^3 - 5x^2 - 3x + 15 = 0$

Solution: $x^3 - 5x^2 - 3x + 15 = 0$

$$x^2(x - 5) - 3(x - 5) = 0$$
$$(x-5)(x^2 - 3) = 0$$

Solving gives: $x - 5 = 0$ → $x = 5$ or $x^2 - 3 = 0$ → $x^2 = 3$ → $x = \pm \sqrt{3}$

And → solution set is $\{5, \pm \sqrt{3}\}$

Practice -5	Solve the quadratic equation by grouping: $x^3 - 2x^2 - 3x + 6 = 0$

4.4 Division of Polynomials

Objectives: 1. Long Division
            2. Synthetic Division

1. Long division: terms are used as they are with variables and coefficients, and powers in descending order, if one order term is missing it should be replaced with a zero, or empty space.

Example -1	Divide the polynomial by binomial:
	$3x^3 + x^2 + 4x - 6$ by x + 3

Solution: to divide $\dfrac{3x^3 + x^2 + 4x - 6}{x + 3}$ we follow these steps:

$$
\begin{array}{r}
3x^2 - 8x + 28 \\
x+3 \overline{\smash{\big)}\, 3x^3 + x^2 + 4x - 6} \\
+ 3x^3 + 9x^2 \quad\longleftarrow \text{Change the sign of this raw} \\
\hline
- 8x^2 + 4x \\
- 8x^2 - 24x \quad\longleftarrow \text{Change the sign of this raw} \\
\hline
+ 28x - 6 \\
+ 28x + 84 \quad\longleftarrow \text{Change the sign of this raw} \\
\hline
- 90 \quad\longleftarrow \text{this is the remainder}
\end{array}
$$

Dividend

Then $\dfrac{3x^3 + x^2 + 4x - 6}{x + 3} = (3x^2 - 8x + 28) + \dfrac{-90}{x + 3}$ Remainder

Quotient   Divisor

Practice -1	Divide the polynomial by binomial:
	$2x^4 - x^3 + 16x^2 - 4$ by 2x - 1

2. Synthetic Division: This method is a short cut for long division, it concentrates on Coefficients only, as before if a power term is missing it should be replaced with a space, a zero, or dashed area.

Example -2	Divide the polynomial by binomial:
	$2x^3 + 4x^2 + 8x - 8$ by $x + 2$

Solution: to divide $\dfrac{3x^3 + 4x^2 + 8x - 8}{x + 2}$ we follow these steps:

$$
\begin{array}{r|rrrr}
-2 & 3 & 4 & 8 & -8 \\
   &   & -6 & 4 & -24 \\
\hline
   & 3 & -2 & 12 & -32
\end{array}
$$

$\longleftarrow$ Remainder

Then $\dfrac{3x^3 + 4x^2 + 8x - 8}{x + 2} = (3x^2 - 2x + 12) + \dfrac{-32}{x + 2}$

Practice -2	Divide the polynomial by binomial:
	$2x^5 - 2x^4 + 3x^2 - x + 1$ by $x - 2$

4.5 Solving Polynomial inequalities

Steps of Solving polynomial Inequalities	Suppose that the polynomial is in one of the following forms: $F(x) < 0$; $F(x) > 0$; $F(x) \le 0$; $F(x) \ge 0$ To solve first locate the zero's of $F(x)$, use the zero's to divide the real numbers line into intervals.

Example -1	Solve the polynomial inequality: $x^3 \le 3x^2$

Solution: $x^3 \le 3x^2$

As mentioned in the steps above the inequality has to be as one of the forms shown above
Or with zero on the right side: Then subtract $3x^2$ from both sides:

$$x^3 - 3x^2 \le 0 \text{ factor this step}$$
$$x^2(x-3) \le 0$$

```
                    –  –   –  –   0 +  +   +  +
       x  ————————————————————+————————————

                    –   –   – – 0 +  +   +  +
       x  ————————————————————+————————————

                    –   – – – – – – 3 +  +
     x–3  ————————————————————+————+—————

                    –   –  – – 0 –  3 + + +
$x^2(x-3) \le 0$  ————————————————————+————+—————
```

Since the solution must be negative, the solution falls between the regions: $(-\infty, 3]$

Practice -1	Solve the polynomial inequality: $x^3 > x^2$

Example -2	Solve the polynomial inequality: $(x-2)(x-3) \le 0$

Solution: $(x–2)(x–3) \le 0$

$$
\begin{array}{c}
\quad - \quad - \quad - \quad - \quad - \; 2 \; + \quad + \quad + \quad + \\
x–2 \quad \rule{6cm}{0.4pt} \\[4pt]
\quad - \quad - \quad - \quad - \quad - \; 3 \; + \quad + \quad + \\
x–3 \quad \rule{6cm}{0.4pt} \\[4pt]
+ \quad + \quad + \quad + \quad + \; 2 \; - \; 3 \; + \quad + \quad + \\
(x–2)(x–3) \le 0 \quad \rule{6cm}{0.4pt}
\end{array}
$$

Since the solution must be negative and closed interval, then the solution falls between the regions: [2, 3]

Practice -2	Solve the polynomial inequality: $3x^2 < -9x$

4.6 Applications of Polynomials

Polynomials can be used to solve a variety of problems in real live, we present some basic ones in this section.

Example -1	Write the polynomial that represents the volume of an open box with length = 16 – x, width = 8–x, and height = x

Solution: Since volume of the box is = length x width x height

Then the volume = (16 –x) (8–x) x

$$\text{Volume} = x^3 - 24x^2 + 128x$$

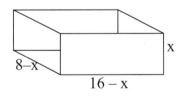

Practice -1	Write the polynomial that represents the volume of an open box with length = 10 – x, width = 4–x, and height = x

Example -2	Write the polynomial that represents the area of the shaded region in the given graph. 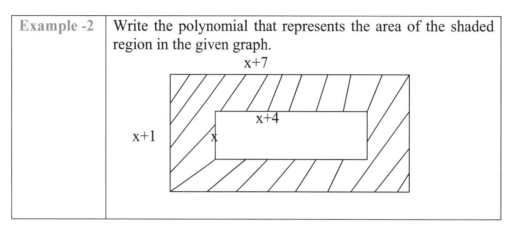

Solution: The shaded area is = Large area – small area

$$= (x+1)\,(x+7) - x(x+4)$$
$$= x^2 + 8x + 7 - x^2 - 4x$$
$$= 4x + 7$$

Practice -1	Express the area of the plane as a polynomial in the standard form: 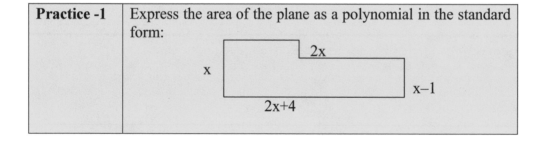

Factor the following polynomials:
1. $4x - 8$
2. $3x^2 - 9xy + 12\ xy^2$

Factor the difference of two squares:
5. $x^2 - 81$
6. $49x^2 - 9$
7. $144x^2 - 36y^2$

Factor the sum of two cubes:
8. $8x^3 + 27$
9. $64 + 125x^3$

Factor the difference of two cubes:
10. $216\ x^3 - 27$
11. $8 - 27x^3$

Factor by grouping:
10. $6x^2 + 9x + 4x + 6$
11. $2x^2 - 4x - x + 2$

Factor completely:
12. $9x^2 - 24x + 16$
13. $5 + 11x - 16x^2$

Use synthetic division to find the quotient and remainder:
14. $3x^3 + 2x^2 - x + 3$ divide by $x-3$
15. $2x^4 - x^3 + 2x - 1$ divide by $x- 1/2$

Use synthetic division and the remainder theorem to find the indicated function value:
16. $f(x) = 4x^3 + 5x^2 - 6x - 4$; $f(-2)$
17. $f(x) = 2x^5 - 3x^4 + x^3 - x^2 + 2x - 1$; $f(2)$

Solve the following polynomials by factoring:

18. $x^2 = 12x - 36$

19. $1/4\, x^2 - 5/2\, x + 6 = 0$

20. $2x(x+3) = 5x + 15$

Solve the polynomial inequalities and graph the solution:

21. $(x-3)(x+1) > 0$

22. $x^2 - 6x + 8 \leq 0$

23. $x^3 \geq 4x^2$

24. A ball is projected in the air with initial velocity of 20ft/s from the ground, its height is
    Modeled as: $y\,(t) = -16\, t^2 + 80t + 120.$

    a. Where would the object be at time t=5 seconds?

    b. At what time will the object be at 40 ft above the ground?

25. Mark wants to fence his backyard which has a rectangular shape of length 3 yards greater
    than the width with area of 180 yard^2. How much material does he need to fence it?

Chapter- 4 Test

1. Factor the following polynomials: $5x - 10$

2. Factor the difference of two squares:
   a. $4x^2 - 49y^2$
   b. $81x^2 - 16$

3. Factor the sum of two cubes:
   a. $27x^3 + 8$
   b. $125 + 8x^3$

4. Factor the difference of two cubes:
       a. $x^3 - 343$
       b. $1 - 27x^3$

5. Factor by grouping:
a. $x^2 + x + 4x + 4$
b. $x^2 - 8x + 16 - y^2$

6. Factor completely:
a. $3x^3 - 15x^2 - 24x$
b. $15x^3 - 25x^2 + 10x$

7. Use synthetic division and the remainder theorem to find the indicated function value:
$$f(x) = 3x^3 - 7x^2 - 2x + 5 \; ; \; f(-3)$$

8. Solve the polynomial inequalities and graph the solution:
$$(x - 3)(x + 2)(x - 5) \geq 0$$

9. A ball is projected in the air with initial velocity of 40ft/s from the ground, its height is
   Modeled as: $y(t) = -16t^2 + 90t + 180$.
       a. Where would the object be at time t=6 seconds?
       b. At what time will the object be at maximum point above the ground?

10. Mark wants to fence his backyard which has a rectangular shape; of length 3 yards greater
    Than twice the width, and has area of 200 yard^2. How much material is required for the
    Fence ?

5. Rational Functions

5. Rational Functions

Objectives:
1. Introduction to Rational Functions
2. Multiplication and Division of Rational Functions
3. Addition and Subtraction of Rational functions
4. Solving Rational Equations
5. Applications

5. 1 Introduction

A rational function R(x) is a fraction form of both of its numerator and denominator are polynomials p(x), and q(x), and can be written in standard form as:

$$R(x) = \frac{P(x)}{q(x)} = \frac{a_n x^n + a_{n-1} x^{n-1} + \ldots + ax + a}{b_m x^m + b_{m-1} x^{m-1} + \ldots + bx + b}, \text{ where } q(x) \neq 0$$

Domain of Rational Functions

Example -1	Find the domain of the rational function: $$F(x) = \frac{x-1}{2x^2 + 5x - 3}$$

Solution: the domain of the rational functions is determined by its nonzero denominator only.

$2x^2 + 5x - 3 \neq 0$ Factor this polynomial

$(2x-1)(x+3) \neq 0 \rightarrow$ then $2x - 1 \neq 0$ or $x+3 \neq 0$

Then the domain of F(x) is $\{x/ \ x \neq -3, x \neq 1/2\}$ or $\{(-\infty, -3) \cup (-3, 1/2) \cup (1/2, \infty)\}$

This can also be expressed verbally as:

The domain of F(x) is all the real numbers except –3, and 1/2.

Practice -1	Find the domain of the rational function: $$F(x) = \dfrac{x + 3}{x^2 - 2x - 15}$$

5. 2 Multiplication and division of Rational Functions

Example -1	Multiply $$\dfrac{x - 1}{x + 1} \cdot \dfrac{x^2 - 1}{2x - 2}$$

Solution:

$$\frac{x - 1}{x + 1} \cdot \frac{x^2 - 1}{2x - 2} = \frac{x - 1}{x + 1} \cdot \frac{(x - 1)(x + 1)}{2(x - 1)} = \frac{x - 1}{2}$$

Practice -1	Multiply $$\dfrac{5x}{x^2 - 6} \cdot \dfrac{x^2 - 10x + 24}{x^2 - x - 12}$$

Example -2	Multiply $$\dfrac{x^3 - 27}{x^2 - 9} \cdot \dfrac{x^2 + 3x}{x^2 + 3x + 9}$$

Solution: $\dfrac{x^3 - 27}{x^2 - 9} \cdot \dfrac{x^2 + 3x}{x^2 + 3x + 9} = \dfrac{(x-3)(x^2 + 3x + 9)}{(x-3)(x+3)} \cdot \dfrac{x(x+3)}{(x^2 + 3x + 9)} = x$

Practice -2	Multiply $\dfrac{x^2 - 4}{9} \cdot \dfrac{3x^2 - 9x}{x^2 - 5x + 6}$

Example -3	Multiply $\dfrac{x^2 + 4}{x^2 - 4} \cdot \dfrac{x^2 + 5x + 6}{x^2 - 1}$

Solution: $\dfrac{x^2 + x}{x^2 - 4} \cdot \dfrac{x^2 + 5x + 6}{x^2 - 1} = \dfrac{x(x+1)}{(x-2)(x+2)} \cdot \dfrac{(x+2)(x+3)}{(x-1)(x+1)}$

$$= \dfrac{x(x+3)}{(x-2)(x-1)}$$

Practice -3	Multiply $\dfrac{x+1}{x^2 - 9} \cdot \dfrac{x^2 + 5x + 6}{x+2}$

Example -4	Multiply $(x^2 - 16) \cdot \dfrac{x-1}{x^2 + 3x - 4}$

Solution: $(x^2 - 16) \cdot \dfrac{x-1}{x^2 + 3x - 4} = (x - 4)(x + 4) \cdot \dfrac{x-1}{(x-1)(x+4)} = (x - 4)$

Practice -4	Multiply $(x^2 - 9) \cdot \dfrac{x-3}{x^2 - 6x + 9}$

Example -5	Divide $(x^2 - 16) \div \dfrac{x+4}{x^2 - 5x + 4}$

Solution: $(x^2 - 16) \div \dfrac{(x+4)(x-1)}{x^2 - 5x + 4} = (x-4)(x+4) \div \dfrac{(x+4)(x-1)}{(x-1)(x-4)}$

$$= (x-4)(x+4) \cdot \dfrac{(x-1)(x-4)}{(x+4)(x-1)}$$

$$= (x - 4)^2$$

Practice -5	Divide
	$(x+1) \div \dfrac{x^2+7x+6}{x+2}$

Solution: $(x+1) \div \dfrac{x^2+7x+6}{x+6} = (x+1) \cdot \dfrac{x+6}{x^2+7x+6}$

$$= (x+1) \cdot \dfrac{(x+6)}{(x+1)(x+6)} = 1$$

Example -6	Divide
	$\dfrac{x-3}{x+7} \div \dfrac{2x-6}{3x+21}$

Solution: $\dfrac{x-}{x+} \div \dfrac{2x-6}{3x+21} = \dfrac{x-3}{x+7} \cdot \dfrac{3(x+7)}{2(x-3)} = 3/2$

Practice -6	Divide
	$\dfrac{x^2-4}{x^3-8} \div \dfrac{2x}{x+2}$

Example -7	Multiply and Divide
	$$\dfrac{2x^3-10x^2}{x^2-8x+15} \cdot \dfrac{3x^3-27x}{2x^2+3x} \div \dfrac{x^2-6x-27}{14x+21}$$

Solution: $\dfrac{2x^3-10x^2}{x^2-8x+15} \cdot \dfrac{3x^3-27x}{2x^2+3x} \div \dfrac{x^2-6x-27}{14x+21}$

First change the division into multiplication:

$$= \dfrac{2x^3-10x^2}{x^2-8x+15} \cdot \dfrac{3x^3-27x}{2x^2+3x} \cdot \dfrac{14x+21}{x^2-6x-27}$$

$$= \dfrac{2x^2\,(x-5)}{(x-5)\,(x-3)} \cdot \dfrac{3x(x^2-9)}{x\,(2x+3)} \cdot \dfrac{7\,(2x+3)}{(x-9)\,(x+3)}$$

$$= \dfrac{2x^2\,(\cancel{x-5})}{(\cancel{x-5})\,(\cancel{x-3})} \cdot \dfrac{3\cancel{x}\,(\cancel{x-3})\,(\cancel{x+3})}{\cancel{x}\,(\cancel{2x+3})} \cdot \dfrac{7\,(\cancel{2x+3})}{(x-9)\,(\cancel{x+3})} = \dfrac{42x^2}{x-9}$$

Practice -7	Multiply and Divide
	$$\dfrac{4x+8}{3x^2-4x-20} \cdot \dfrac{x^2+5x}{2x^2+11x+5} \div \dfrac{4x}{6x^2-17x-10}$$

5. 3 Addition and Subtraction of Rational Functions

Rules of Addition and Subtraction of Rational functions	Adding two rational functions with the same denominator $$\frac{P_1}{q} + \frac{P_2}{q} = \frac{P_1 + P_2}{q}$$ Subtracting two rational functions with the same denominator $$\frac{P_1}{q} - \frac{P_2}{q} = \frac{P_1 + P_2}{q}$$

Example -1	Add and Simplify $$\frac{2x-3}{x} + \frac{5x-1}{x}$$

Solution: $$\frac{2x-3}{x} + \frac{5x-1}{x} = \frac{2x-3+5x-1}{x} = \frac{7x-4}{x}$$

Practice -1	Add and Simplify $$\frac{x}{x+2} + \frac{2}{x+2}$$

Example -2	Add and Simplify
	$$\dfrac{x^2-2x}{x^2+2x} + \dfrac{x^2+x}{x^2+2x}$$

Solution: $\dfrac{x^2-2x}{x^2+2x} + \dfrac{x^2+x}{x^2+2x} = \dfrac{x^2-2x+x^2+x}{x^2+2x} = \dfrac{2x^2-x}{x^2+2x} = \dfrac{x(2x-1)}{x(x+2)} = \dfrac{2x-1}{x+2}$

Practice -2	Subtract and Simplify
	$$\dfrac{4x}{5x-3} - \dfrac{3x-1}{5x-3}$$

Example -3	Subtract and Simplify
	$$\dfrac{x^2+2x-6}{x^2-5x+4} - \dfrac{4x^2-4-2x^2}{x^2-5x+4}$$

Solution: $\dfrac{x^2+2x-6}{x^2-5x+4} - \dfrac{4x-4-2x^2}{x^2-5x+4} = \dfrac{x^2+2x-6-4x+4+2x^2}{x^2-5x+4}$

$$= \dfrac{3x^2-2x-2}{x^2-5x+4}$$

Practice-3	Subtract and Simplify
	$$\dfrac{x^2+2x}{x^2-4x+3} - \dfrac{-4-2x^2}{x^2-4x+3}$$

Example -4	Add and Simplify
	$$\dfrac{2}{x-1} + \dfrac{10}{x-2}$$

Solution: Since the denominators are different, then we have to find the least common denominator LCD = (x-1)(x-2), then we make the denominator the same,

$$\dfrac{2(x-2)}{(x-1)(x-2)} + \dfrac{10(x-1)}{(x-2)(x-1)} \text{ Now we add two rational of the same denominator}$$

$$= \dfrac{2(x-2)+10(x-1)}{(x-2)(x-1)} = \dfrac{2x-4+10x-10}{(x-2)(x-1)} = \dfrac{12x-14}{(x-1)(x-2)}$$

Practice -4	Add and Simplify
	$$\dfrac{13}{x+1} + \dfrac{11}{x-2}$$

Example -5	Subtract and Simplify
	$$\dfrac{x-1}{x^2-4x+3} - \dfrac{x+2}{x^2-x-2}$$

Solution: $\dfrac{x-1}{x^2-4x+3} - \dfrac{x+2}{x^2-x-2} = \dfrac{x-1}{(x-1)(x-3)} - \dfrac{x+2}{(x-2)(x+1)}$

The LCD = $(x-1)(x-3)(x-2)(x+1)$

$$= \frac{(x-1)(x-2)(x+1)}{(x-1)(x-3)(x-2)(x+1)} - \frac{(x+2)(x-1)(x-3)}{(x-2)(x+1)(x-1)(x-3)}$$

$$= \frac{(x-1)(x-2)(x+1) - (x+2)(x-1)(x-3)}{(x-2)(x+1)(x-1)(x-3)}$$

$$= \frac{(x-1)[(x-2)(x+1) - (x+2)(x-3)]}{(x-2)(x+1)(x-1)(x-3)}$$

$$= \frac{(x-1)[x^2-x-2 - x^2+x+6]}{(x-2)(x+1)(x-1)(x-3)}$$

$$= \frac{4(x-1)}{(x-2)(x+1)(x-1)(x-3)} = \frac{4}{(x-2)(x+1)(x-3)}$$

Example -6	Add and Subtract
	$\dfrac{2x}{x-2} + \dfrac{8}{x^2-4} - \dfrac{x}{x+2}$

Solution: $\dfrac{2x}{x-2} + \dfrac{8}{x^2-4} - \dfrac{x}{x+2}$

$$= \dfrac{2x}{x-2} + \dfrac{8}{(x-2)(x+2)} - \dfrac{x}{x+2} \quad \text{The LCD} = (x-2)(x+2)$$

$$= \dfrac{2x(x+2)}{(x-2)(x+2)} + \dfrac{8}{(x-2)(x+2)} - \dfrac{x(x-2)}{(x-2)(x+2)}$$

$$= \dfrac{2x(x+2) + 8 - x(x-2)}{(x-2)(x+2)} = \dfrac{2x^2 + 4x + 8 - x^2 + 2x}{(x-2)(x+2)} = \dfrac{x^2 + 6x + 8}{(x-2)(x+2)}$$

$$= \dfrac{(x+4)(x+2)}{(x-2)(x+2)} = \dfrac{x+4}{x-2}$$

Practice-6	Add and Subtract
	$\dfrac{2x-1}{x-3} + \dfrac{x+2}{x^2-2x-3} - \dfrac{4}{x+1}$

5. 4 Solving Rational Equations

Example -1	Solve the Rational Equation $$\frac{x+1}{4} = x-1$$

Solution: Multiply the equation by 4 to get rid of the fraction

$$\cancel{4} \cdot \frac{x+1}{\cancel{4}} = 4\,(x-1)$$

$x+1 = 4x - 4$ → the solution set is $\{x = 5/3\}$

Practice -1	Solve the Rational Equation $$\frac{5}{x-2} = \frac{4}{x+1}$$

Example -2	Solve the Rational Equation $$\frac{x+1}{x-2} - \frac{x+3}{x} = \frac{6}{x^2 - 2x}$$

Solution: $$\frac{x+1}{x-2} - \frac{x+3}{x} = \frac{6}{x^2 - 2x}$$

$$\frac{x+1}{x-2} - \frac{x+3}{x} = \frac{6}{x(x-2)} \qquad \text{LCD} = x(x-2)$$

Multiply the equation by LCD = x(x–2) to get rid of the fractions

$$x(x-2) \frac{x+1}{x-2} - x(x-2) \frac{x+3}{x} = x(x-2) \frac{6}{x(x-2)}$$

$x(x+1) - (x-2)(x+3) = 6$

$x^2 + x - x^2 - x + 6 = 6$

$6 = 6$ or $0 = 0$ is a true stamen which means the equation has many solutions except

0 and 2 which makes the equation undefined.

Practice -2	Solve the Rational Equation $$\frac{3}{x+1} - \frac{4}{x+5} = \frac{2}{x^2 + 6x + 5}$$

5. 5 Solving Rational Inequalities

This is similar what was done in polynomial inequalities

Example -1	Solve the Rational Inequality $$\frac{x+2}{x-1} < 0$$

Solution:

x+ 3

$$- \quad - \quad -3 + \quad + \quad + \quad + \quad + \quad + \quad +$$

x–1

$$- \quad - \quad - \quad - \quad - \quad - \quad 1 + \quad + \quad + \quad +$$

$$\dfrac{x+3}{x-1} < 0$$

$$+ \quad -3 \quad - \quad 1 \quad +$$

Then the solution set is $\{(-3, 1)\}$

Practice -1	Solve the Rational Inequality $$\dfrac{3x+1}{x-2} \geq 0$$

Example -2	Solve the Rational Inequality $$\dfrac{x+2}{x-5} > \dfrac{3}{x-5}$$

Solution:

$$\dfrac{x+2}{x-5} - \dfrac{3}{x-5} > 0$$

$\dfrac{x-1}{x-5} > 0$ solving in the same manner, we find that the positive solution exists in the region in the region $\rightarrow \{(-\infty, 1) \cup (5, \infty)\}$

Practice -2	Solve the Rational Inequality $$\frac{4x+2}{x+4} \geq 1$$

5. 6 Applications

Example -1	Find the area of a rectangle with length$= \dfrac{2x}{(x+5)}$, And width $= \dfrac{x}{(x+3)}$.

Solution: Area of rectangle = Length • Width

$$= \frac{2x}{(x+3)} \bullet \frac{x}{(x+5)} = \frac{2x^2}{(x+3)(x+5)}$$

Practice -1	Find the area of a rectangle with length$= \dfrac{3x}{(x+1)}$, And width $= \dfrac{x}{(x+2)}$.

Example -2	Given: $$f(x) = \frac{x}{x+2} \quad \text{and} \quad g(x) = \frac{1}{x^2 + 2x + 1}$$ Find: a) $(fg)(x)$ b) $(f/g)(x)$ c) $(f+g)(x)$ d) $(f-g)(x)$

Solution: a) $(fg)(x) = f(x) \cdot g(x) = \dfrac{x}{(x+2)(x^2+2x+1)}$

b) $(f/g)(x) = f(x) / g(x) = \dfrac{x(x^2+2x+1)}{(x+2)}$

c) $(f+g)(x) = f(x) + g(x) = \dfrac{x}{x+2} + \dfrac{1}{x^2+2x+1}$

$$= \dfrac{x(x^2+2x+1) + x+2}{(x+2)(x^2+2x+1)}$$

$$= \dfrac{x^3+2x^2+x+x+2}{(x+2)(x^2+2x+1)}$$

$$= \dfrac{x^3+2x^2+2x+2}{(x+2)(x^2+2x+1)}$$

d) $(f-g)(x) = f(x) - g(x) = \dfrac{x}{x+2} - \dfrac{1}{x^2+2x+1}$

$$= \dfrac{x^3+2x^2-2}{(x+2)(x^2+2x+1)}$$

Practice -2	Given: $f(x) = \dfrac{x}{x+2}$ and $g(x) = \dfrac{2}{x+1}$ Find: a) $(fg)(x)$ b) $(f/g)(x)$ c) $(f+g)(x)$ d) $(f-g)(x)$

Chapter-5 Exercise

Multiply the following rational expressions:

1. $(x + 1) \cdot \dfrac{x + 2}{x^2 + 7x + 6}$

2. $\dfrac{x^2 - 4}{x^3 - 8} \cdot \dfrac{2x}{x + 2}$

3. $\dfrac{x^2 - 9}{x^2 - x - 6} \cdot \dfrac{x^2 + 5x + 6}{x^2 + x - 6}$

Divide the following rational expressions:

4. $\dfrac{x^2 - x}{15} \div \dfrac{x - 1}{5}$

5. $\dfrac{x^2 - 25}{2x - 2} \div \dfrac{x^2 + 10x + 25}{x^2 + 4x - 5}$

6. $\dfrac{x^2 - 4y^2}{x^2 + 3xy + 2y^2} \div \dfrac{x^2 - 4xy + 4y^2}{x + y}$

Add the following rational expressions:

7. $\dfrac{20}{15x} + \dfrac{5}{15x}$

8. $\dfrac{x^2 + 2}{x^2 - 6x - 7} + \dfrac{-23 - 4x}{x^2 + 6x - 7}$

9. $\dfrac{4}{x - 2} + \dfrac{3}{x + 1}$

Subtract the following rational expressions:

10. $\dfrac{3x}{7x - 4} - \dfrac{2x - 1}{7x - 4}$

11. $\dfrac{x^2 + 6x + 2}{x^2 + x - 6} - \dfrac{2x - 1}{x^2 + x - 6}$

12. $\dfrac{3x}{x^2 - 25} - \dfrac{8}{x + 5}$

Solve the following rational equations:

13. $\dfrac{8}{x} + \dfrac{1}{4} = \dfrac{4}{2x}$

14. $\dfrac{8}{x+2} = \dfrac{2}{x-2}$

15. $x - \dfrac{6}{x} = -1$

Solve the following rational inequalities and graph the solution:

16. $\dfrac{x-4}{x+2} \leq \mathbf{0}$

17. $\dfrac{x^2 - 3x + 2}{x^2 - 2x - 3} \geq 0$

18. $\dfrac{3}{x+2} > \dfrac{2}{x-2}$

A company manufactures bicycles for kids. The average cost function of their production Was modeled as:

$$C(x) = \frac{6000 + 300x}{x}$$

Where, C(x) is the average cost of production in dollars, and x is the number of bicycles Produced Per month:

19. What is the average cost of production of 100 bicycles per month?

20. How many bicycles should the company produce so that the cost does not exceed $350?

Chapter-5 Test

1. Multiply and divide the rational expressions:

$$\frac{10x^2 + 13x - 3}{3x^2 - 8x + 5} \cdot \frac{2x^2 - 3x - 2x + 3}{25x^2 - 10x + 1} \div \frac{4x^2 - 9}{15x^2 - 28x + 5}$$

2. Add and subtract the rational expressions:

$$\frac{3x}{x^2 - 4} + \frac{5x}{x^2 + x - 2} - \frac{3}{x^2 - 4x + 4}$$

3. Solve the rational equation:

$$\frac{36 + 4x^2}{x^2 - 9} = \frac{4x}{x + 3} - \frac{12}{x - 3}$$

4. Solve the rational inequality:

$$\frac{1}{4(x-1)} \ge \frac{4}{(x+3)}$$

5. Solve the rational inequality

$$\frac{3x+6}{x^2-6x+5} \ge 0$$

6. Radical Functions

6. Radical Functions

6.1 Introduction

Rules of Radicals	
$x^2 = a \rightarrow$ then $x = \pm\sqrt{a}$	$\sqrt[n]{x^n} = x$ if n is odd and $\lvert x \rvert$ if n is even
$\sqrt{xy} = \sqrt{x} \cdot \sqrt{y}$	$\sqrt{(-x)^2} = \lvert -x \rvert = x$
$\sqrt{x} = x^{1/2}$	$\sqrt{x^2} = \lvert x \rvert = x$
$\sqrt[m]{x^n} = x^{n/m}$	If $\sqrt[3]{x} = y \rightarrow$ then $x = y^3$

6.2 Radical Expressions

Example -1	Simplify:
	a. $\sqrt{81} = 9$
	b. $\sqrt[6]{(x-1)^6} = \lvert x-1 \rvert$
	c. $\sqrt[5]{(x-1)^5} = x-1$

Practice -1	Simplify:
	a. $\sqrt{121}$
	b. $\sqrt[8]{(x-3)^8}$
	c. $\sqrt[7]{(x-3)^7}$

Example -2	Simplify:
	a. $\sqrt{(-3)^2} = |-3| = 3$
	b. $\sqrt{(x+1)^2} = |x+1|$
	c. $\sqrt{49x^6} = \sqrt{(7x^3)^2} = |7x^3|$
	d. $\sqrt{x^2-2x+1} = \sqrt{(x-1)^2} = |x-1|$

Domain of Radical Functions

$F(x) = \sqrt{g(x)}$ → the domain of F(x) is $g(x) \geq 0$

Example -3	Find the domain of
	$F(x) = \sqrt{x-6}$
	$x-6 \geq 0$ → then domain of F(x) is $x \geq 6$ or $\{ [6,\infty) \}$

Practice -3	Find the domain of $F(x) = \sqrt{3x-9}$

Example -4	Simplify the following expressions: a. $7^{1/2} \cdot 7^{3/4} = 7^{1/2+3/4} = 7^{5/4}$ b. $\dfrac{49\,x^{1/2}}{7\,x^{3/4}} = 7\,x^{1/2-3/4} = 7x^{-1/4}$ c. $(5^{3/5})^{5/2} = 5^{3/5 \cdot 5/2} = 5^{3/2}$ d. $(x^{-1/6} \cdot y^{1/3})^2 \; (x^{-1/6})^2 \cdot (y^{1/3})^2 = x^{-2/6}\,y^{2/3} = x^{-1/3}\,y^{2/3}$

6.3 Multiplying and Simplifying Radical expressions

Example -1	Multiply the following radicals: a. $\sqrt{5}\,\sqrt{7} = \sqrt{5.7} = \sqrt{35}$ b. $\sqrt[3]{11}\;\sqrt[3]{3} = \sqrt[3]{11(3)} = \sqrt[3]{33}$ c. $\sqrt{x+1} \cdot \sqrt{x-1} = \sqrt{(x+1)(x-1)} = \sqrt{x^2-1}$ d. $\sqrt{x+1} \cdot \sqrt{x+1} = \sqrt{(x+1)^2} = |x+1|$

6.4 Adding and Subtracting Radical expressions

Example -1	Add the following radicals:
	$\sqrt{2} + \sqrt{18} = \sqrt{2} + \sqrt{2 \cdot 9}$
	$\qquad\qquad = \sqrt{2} + \sqrt{2}\,\sqrt{9}$
	$\qquad\qquad = \sqrt{2} + \sqrt{2} \cdot 3 = \sqrt{2}\,(1+3) = 4\sqrt{2}$

Practice -1	Add the following radicals:
	$\sqrt{20} + 5\sqrt{45}$

Example -2	Subtract the following radicals:
	$3\sqrt{27x} - 5\sqrt{12x} = 3\sqrt{3x \cdot 9} - 5\sqrt{3x \cdot 4}$
	$\qquad\qquad\qquad = 3\sqrt{3x}\,\sqrt{9} - 5\sqrt{3x}\,\sqrt{4}$
	$\qquad\qquad\qquad = 3\sqrt{3x} \cdot 3 - 5\sqrt{3x} \cdot 2$
	$\qquad\qquad\qquad = 9\sqrt{3x} - 10\sqrt{3x}$
	$\qquad\qquad\qquad = \sqrt{3x}\,(9-10)$
	$\qquad\qquad\qquad = -\sqrt{3x}$

Practice -2	Subtract the following radicals:
	$3\sqrt{12x} - 6\sqrt{27x} = 3\sqrt{3 \cdot 3x} - 6\sqrt{9 \cdot 3x}$
	$\qquad\qquad\qquad = 3\sqrt{4}\,\sqrt{3x} - 6\sqrt{9}\,\sqrt{3x}$
	$\qquad\qquad\qquad = 6\sqrt{3x} - 18\sqrt{3x}$
	$\qquad\qquad\qquad = \sqrt{3x}\,(6-18)$
	$\qquad\qquad\qquad = -12\sqrt{3x}$

Example - 3	Subtract the following radicals:
	$$3\sqrt[3]{24} - 5\sqrt[3]{81} = 3\sqrt[3]{3.2^3} - 5\sqrt[3]{3.3^3}$$ $$= 3.2\sqrt[3]{3} - 5.3\sqrt[3]{3}$$ $$= 6\sqrt[3]{3} - 15\sqrt[3]{3}$$ $$= (6-15)\sqrt[3]{3}$$ $$= -9\sqrt[3]{3}$$

Practice - 3	Subtract the following radicals:
	$$15\sqrt[3]{81} - 3\sqrt[3]{24}$$

Rationalizing Radical denominators

Example - 4	Rationalize the Denominator
	$$\frac{\sqrt{3}}{\sqrt{6}} = \frac{\sqrt{3}}{\sqrt{6}} \cdot \frac{\sqrt{6}}{\sqrt{6}}$$ $$= \frac{\sqrt{3.6}}{\sqrt{6.6}} = \frac{\sqrt{18}}{6}$$

Practice - 4	Rationalize the Denominator $$\frac{\sqrt{3}}{\sqrt{7}}$$

Example - 5	Rationalize the Denominator $$\sqrt[3]{\frac{7}{16}} = \sqrt[3]{\frac{7}{4^2}} = \sqrt[3]{\frac{7\cdot4}{4^2\cdot4}} = \sqrt[3]{\frac{28}{4^3}} = \frac{\sqrt[3]{28}}{4}$$

Practice - 5	Rationalize the Denominator $$\sqrt[3]{\frac{x}{9y}}$$

Example - 6	Rationalize the Denominator $$\frac{3+\sqrt7}{\sqrt5-\sqrt2} = \frac{3+\sqrt7}{\sqrt5-\sqrt2}\cdot\frac{\sqrt5+\sqrt2}{\sqrt5+\sqrt2} = \frac{(3+\sqrt7)(\sqrt5+\sqrt2)}{(\sqrt5-\sqrt2)(\sqrt5+\sqrt2)}$$ $$= \frac{(3+\sqrt7)(\sqrt5+\sqrt2)}{5-2} = \frac{(3+\sqrt7)(\sqrt5+\sqrt2)}{3}$$

Practice - 6	Rationalize the Denominator $$\frac{18}{2\sqrt3+3}$$

6.5 Solving Radical Equations

Example - 1	Solve the radical equation: $$\sqrt{3x-5} = 10$$

Solution: $\sqrt{3x-5} = 10$

Square both sides to get rid of the square root

$(\sqrt{3x-5})^2 = (10)^2$

$3x - 5 = 100$ solving for x gives $\rightarrow$ $x = \dfrac{105}{3} = 35$

Practice - 1	Solve the radical equation:
	$\sqrt{2x+8} = x$

Example - 2	Solve the radical equation
	$\sqrt{6x+2} = \sqrt{5x+3}$

Solution: $\sqrt{6x+2} = \sqrt{5x+3}$

Square both sides to get rid of the square root

$(\sqrt{6x+2})^2 = (\sqrt{5x+3})^2$

$6x + 2 = 5x + 3$ Solving for x gives $\rightarrow$ $x = 1$

6.6 Complex Numbers

Rules of Radicals	For $a > 0$, $\sqrt[n]{a}$ is a real number for even n and odd n. For $a < 0$, $\sqrt[n]{a}$ is a real number for odd n, and complex number for even n $\sqrt{-1} = i$ $\sqrt{(-1)^2} = -1$

Example	a. $\sqrt{-5} = \sqrt{5}\,i$ b. $\sqrt{-25} = 5\,i$ c. $\sqrt{-8} = 2\sqrt{2}\,i$

6.7 Applications

To solve problems in real life scientists collect data and form a mathematical model for the subject under the study, then use the model to answer questions related to the topic. The following model was created for the heights of boys for different ages:

$$f(x) = 3\sqrt{x} + 20$$

Where, f(x) is the median height in inches, and x is the age in months. Use the model to answer the following questions:

1. What is the median height for the boys of 40 months old?
2. What age group falls in the median height of 30 inches?
3. Find age corresponding to the median height larger than 26 inches.

Solution: 1. to find the median height for boys age 40 months, replace x with 40

In the formula:
$$f(x) = 3\sqrt{x} + 20$$
$$f(40) = 3\sqrt{40} + 20 \approx 26.3 \text{ inch}$$

2. To find the age group, replace f(x) with 30 inch to get x:

$$f(x) = 3\sqrt{x} + 20$$
$$30 = 3\sqrt{x} + 20, \text{ then solve for x}$$
$$-20 \qquad -20$$

$$10 = 3\sqrt{x} \rightarrow \sqrt{x} = 10/3 \rightarrow \text{ then } x = (10/3)^2 \approx 11 \text{ months.}$$

3. If f(x) > 26 inch what is x? The problem is to solve the inequality:

$$3\sqrt{x} + 20 > 26$$
$$-20 \quad -20$$

$$3\sqrt{x} \qquad > 6 \rightarrow \sqrt{x} > 2 \rightarrow \text{ or } x > 4 \text{ months}$$

Chapter-6 Exercise

1. Evaluate the following expressions:

a) $\sqrt{\dfrac{9}{16}}$

b) $\sqrt{25 - 16}$

c) $\sqrt{144 + 25}$

2. Graph the function $f(x) = \sqrt{x - 3}$, and give the domain of the function.

3. Evaluate the following radicals:

a) $\sqrt[3]{\dfrac{27}{1000}}$

b) $\sqrt[5]{-1}$

c) $\sqrt[7]{y^7}$

d) $\sqrt[3]{\dfrac{1}{1000}}$

e) $\sqrt[6]{(x+5)^6}$

4. Graph the function $f(x) = \dfrac{\sqrt{x-2}}{\sqrt{7-x}}$ and give the domain of the function.

5. Solve the radical equation: $\sqrt{3x-2} = 8$

6. Solve the radical equation: $x - 2\sqrt{x-3} = 3$

7. Solve the radical equation: $\sqrt{5x+1} = x+1$

8. Solve the radical equation: $\sqrt[3]{2x+3} + 4 = 6$

9. Solve. $\sqrt{3x+6} + 6 = 0$

10. Solve. $\sqrt{5x+5} - 15 = 0$

Chapter-6 Test

1. Evaluate the radical expression:

$$\sqrt{\dfrac{81}{9}} - \sqrt{4}$$

2. Graph the function $f(x) = \sqrt{2x-6}$ and find the domain of the function.

3. Rationalize the denominator $\dfrac{3}{\sqrt{x-5}}$

4. Solve the radical equation $\sqrt{2x-6} = \sqrt{x+1}$

5. Add and subtract the radicals $5\sqrt[3]{81} + 3\sqrt[3]{192} - 8\sqrt[3]{24}$

7. Quadratic Equations

7. Quadratic Equations

> **Objectives:** 1. Introduction
> 2. Solving Quadratic Equations
> 3. Solving using the Square Root Method.
> 4. Solving Using the Completing the Square Method
> 5. Solving using the Quadratic Formula
> 6. Applications.

7.1 Introduction

The quadratic equation is a second degree polynomial with the standard form as,

$$ax^2 + bx + c = 0, \text{ with } a \neq 0$$

The graph of quadratic equation is a parabola, which is opened up if leading coefficient $a > 0$, or opened down when $a < 0$.

Leading Coefficient a	If $a < 0$, then parabola is opened down
	If $a > 0$, the Parabola is opened down

7. 2 Solving Quadratic Equations

Methods of Solving Quadratic equations	Quadratic equations can be solved by:
	1. Factoring
	2. Square root method
	3. Completing the square
	4. Quadratic Formula

7. 3 Solving using the Square Root Method.

Example - 1	Solve the quadratic equation by square root method: $5x^2 = 40$

Solution: $5x^2 = 40$

$$x^2 = \frac{40}{5} = 8$$

$x^2 = 8$, then $x = \pm\sqrt{8} = \pm 2\sqrt{2}$

Practice - 1	Solve the quadratic equation by square root method: $9x^2 = 81$

Example - 2	Solve the quadratic equation by square root method: $7x^2 - 49 = 0$

Solution: $7x^2 - 49 = 0$

$$7x^2 = 49$$

$$x^2 = \frac{49}{7} = 7$$

$x^2 = 7$, then $x = \pm\sqrt{7}$

Practice - 2	Solve the quadratic equation by square root method: $5x^2 - 125 = 0$

Example - 3	Solve the quadratic equation by square root method: $(x+2)^2 = 5$

Solution: $(x+2)^2 = 15$ Take the square root of both sides:

$$\sqrt{(x+2)^2} = \pm\sqrt{5}$$

$x+2 = \pm\sqrt{5}$

$x = -2 \pm \sqrt{5}$ → Solution set is $\{-2 \pm \sqrt{5}\}$

Practice - 3	Solve the quadratic equation by square root method: $(x-3)^2 = 49$

Example - 4	Solve the quadratic equation by square root method: $(x+1/2)^2 = 3/7$

Solution: $(x+1/2)^2 = 3/7$, take the square root of both sides:

$$\sqrt{(x+\tfrac{1}{2})^2} = \pm\sqrt{\frac{3}{7}}$$

$$x + 1/2 = \pm\sqrt{\frac{3}{7}}$$

$$x = -1/2 \pm \sqrt{\frac{3}{7}}$$

Practice - 4	Solve the quadratic equation by square root method: $(x-1/3)^2 = 5/4$

7. 4   Solving Using Completing the Square Method

To solve the quadratic equation, in the standard form, using the method of completing the square, we write the equation in the following form:

$$x^2 + bx + c = 0$$
$$\underline{\quad\quad -c \quad -c \quad}$$

$x^2 + bx = -c$, the left side of this equation will be a complete square if we add to it

The term $(+b/2)^2$

$x^2 + bx + (+b/2)^2 = -c + (+b/2)^2$ this can be written as:

$(x + b/2)^2 = (b/2)^2 - c$ Now the square is complete.

Example - 5	Solve by completing the square method: $x^2 + 4x = 0$

Solution: The term that completes the square is $(+4/2)^2 = (+2)^2$

$$x^2 + 4x + (+2)^2 = 0 + (+2)^2$$
$$(x+2)^2 = 4 \text{ Now we take the square root of both sides}$$

$$\sqrt{(x+2)^2} = \pm\sqrt{4}$$

$x + 2 = \pm 2$

Solution is: $x + 2 = 2$ or x=0, and $x+2 = -2 = -4$

Then the solution set is $\{-4, 0\}$

Practice - 5	Solve by completing the square method: $x^2 - 9x = 0$

Practice-6	Add the term that completes the square for the following:
	a. $x^2 + 20x$
	b. $x^2 + 5x$
	c. $x^2 - 8x$
	d. $x^2 + 5/4x$
	e. $x^2 - 7/3x$
	f. $x^2 + 16x$
	g. $x^2 - 24x$

Example - 6	Solve by completing the square method: $x^2 - 5x + 4 = 0$

Solution: $x^2 - 5x + 4 = 0$ Move the constant term to the right side

$x^2 - 5x = -4$ add the term that completes the L.S $= (-5/2)^2$ to both sides

$x^2 - 5x + (-5/2)^2 = -4 + (-5/2)^2$ and rewrite as:

$(x - 5/2)^2 = -4 + (-5/2)^2$

$(x - 5/2)^2 = -4 + 25/4$

$(x - 5/2)^2 = \frac{25}{4} - 4$

$(x - 5/2)^2 = 9/4$

$\sqrt{(x - \frac{5}{2})^2} = \pm \sqrt{\frac{9}{4}}$ $\rightarrow$ $x - \frac{5}{2} = \pm \frac{3}{2} \rightarrow x = \frac{5}{2} \pm \frac{3}{2}$ Solution set is: {1, 4}

Practice - 7	Solve by completing the square method: $x^2 + 3x - 1 = 0$

7.5 Solving using the Quadratic Formula

Quadratic Formula	Knowing the coefficients (a, b, c) of quadratic equation in the standard form: $ax^2 + bx + c = 0$ we can plug it into the formula to find the solution : $$x = \frac{-b \pm \sqrt{b^2 - 4ac}}{2a}$$
Type of Solutions	Knowing the coefficients (a, b, c) of quadratic equation in the standard form: $ax^2 + bx + c = 0$ we can plug it into the descreminant : $b^2 - 4ac$ and find the type of solution we get as follows: • If $b^2 - 4ac > 0$, we expect to get two distinct and real solutions. • If $b^2 - 4ac < 0$, we expect to get two distinct but not- real solutions. • If $b^2 - 4ac = 0$, we expect to get two equal and real solutions.

Example - 1	For each quadratic equation use the discriminant to find the type of solution to expect, the solve using the quadratic formula: a. $3x^2 + 3x - 4 = 0$ b. $3x^2 - 5x + 7 = 0$ c. $4x^2 - 4x + 1 = 0$

Solution: a) $b^2 - 4ac = (+3)^2 - 4(3)(-4) = 57$ Positive value means we expect to get Two different and real solutions, using quadratic formula we can find them:

$$x = \frac{-b \pm \sqrt{b^2 - 4ac}}{2a} = \frac{-3 \pm \sqrt{57}}{2(3)} = \frac{-3 \pm \sqrt{57}}{6} \text{ is the solution.}$$

b) $b^2 - 4ac = (-5)^2 - 4(3)(7) = -59$ Negative value means we expect two different but not real solutions (complex solutions). Using quadratic formula we can find them:

$$x = \frac{-b \pm \sqrt{b^2 - 4ac}}{2a} = \frac{-(-5) \pm \sqrt{-59}}{2(3)} = \frac{+5 \pm \sqrt{-59}}{6} = \frac{5 \pm i\sqrt{59}}{6} \text{ Complex Solution}$$

c) $b^2 - 4ac = (-4)^2 - 4(4)(1) = 0$ this means we expect two equal (double) solutions.

$$x = \frac{-b \pm \sqrt{b^2 - 4ac}}{2a} = \frac{-(-4) \pm 0}{2(4)} = \frac{4}{8} = 1/2 \text{ Solution set is: } \{1/2, 1/2\}$$

Practice - 1	For each quadratic equation use the discriminant to find the type of solution to expect, the solve using the quadratic formula: a. $2x^2 + 2x - 3 = 0$ b. $5x^2 - 7x + 9 = 0$ c. $x^2 - 2x + 1 = 0$

Solving Quadratic Equations Graphically

The quadratic equation has a shape of a parabola, opened up or down depending on the sign of the leading coefficient if positive, then it will open up, if negative then it will open down.
The solution can be found by the intersection of the parabola with x-axis or the intercepts, if the parabola does not intersect the x-axis that is when the solution is not real, but complex, as shown on the given graphs:

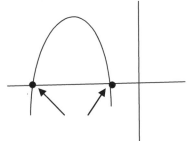

a. Parabola with two real
And positive solutions.

b. Parabola with two real
And negative solutions.

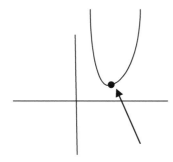

c. Parabola not crossing the x-axis
   means complex solution

d. Parabola touching the x-axis means
   double solutions.

7.6 Applications

Newton's laws of force are used in solving Projectile' Problems in physics which leads into quadratic equations.

Example-1	A ball is projected in the air at an angle of $35°$, its path was modeled by the following equation: $F(x) = -0.01\,x^2 + 0.6\,x + 6,$ Where, x= the horizontal distance (in feet) for the ball in the air $F(x)$ = the height in feet. Show the path of the ball in the air

Solution: Using TI-83 first we set up the window then graph to get the path.

Window
Xmin=0
Xmax=80
Xscl=10
Ymin=0
Ymax=40
Yscl=10

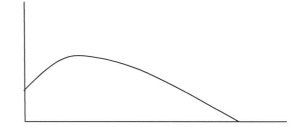

Quadratic equations can be formed by applying Pythagoras theorem to a right triangle.

Example-2	In the right triangle: Find x 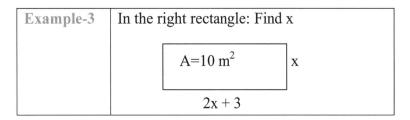

Solution: Applying Pythagorean Theorem:

$$6^2 = (x-2)^2 + x^2$$

$$36 = x^2 - 4x + 4 + x^2$$

The quadratic equation is: $2x^2 - 4x - 32 = 0$
The discriminant gives: $b^2 - 4ac = (-4)^2 - 4(2)(-32) = 272$
And the quadratic formula gives:

$$x = \frac{-b \pm \sqrt{b^2 - 4ac}}{2a} = \frac{-(-4) \pm \sqrt{272}}{2(2)} = \frac{4 \pm \sqrt{272}}{4}$$

Solution set is: $\{-3.1, 5.1\}$

Example-3	In the right rectangle: Find x A=10 m² x 2x + 3

Solution: Area of rectangle = length x width
Then $10 = x(2x+3)$
$\quad\quad 10 = 2x^2 + 3x$ or the quadratic equation $\rightarrow 2x^2 + 3x - 10 = 0$
The discriminant gives: $b^2 - 4ac = (+3)^2 - 4(2)(-10) = 89$

And the quadratic formula gives:

$$x = \frac{-b \pm \sqrt{b^2 - 4ac}}{2a} = \frac{-(3) \pm \sqrt{89}}{2(2)} = \frac{-3 \pm \sqrt{89}}{4}$$

Solution set is: $\{-3.1, 1.6\}$

Chapter – 7 Exercise

Solve the following equations using the square root method:
1. $x^2 = 49$
2. $(x+1)^2 = 9$
3. $(3x+2)^2 = 16$

Solve the equations by completing the square method:
4. $x^2 - 6x + 4 = 0$
5. $2x^2 - 5x - 7 = 0$
6. $1/3\ x^2 - x = 5/3$

Solve the equations using the quadratic formula:
7. $2x^2 = 1 - 4x$
8. $5x^2 - 3x = 1$
9. $2x^2 + x = 7$
10. $\dfrac{x}{x-1} + \dfrac{2}{x+2} = \dfrac{7x+1}{x^2 + x - 2}$

1. Solve the following equation by factoring:
   $X^2 + 4x -21 = 0$
2. Solve the following equation using the square root method:
   $(3x–1)^2 = 4$
3. Solve the equation by completing the square method:
   $2x^2 – 5x + 5 = 0$
4. Solve the equation using the quadratic formula:
   $2x^2 – 3x + 4 = 0$
5. Use the discriminant to determine the type of solution:
   $3x^2 –3x – 7 = 0$

8. Exponential and Logarithmic Functions

8. Exponential and Logarithmic Functions

8.1   Introduction

Logarithmic Functions	The logarithm function of x to base b is written in the form: $\log_b x$ $y = \log_b x \rightarrow$ IFF $x = b^y$
Exponential Functions	An exponential function is a function that can be expressed in the form, $f(x) = b^x$, $b > 0$, $b \neq 0$ b is the base of the exponential function, and base of the logarithmic function too. Logarithmic functions and exponential functions are inverse of each other: $y = \log_b x \longrightarrow x = b^y$

The graph below shows how the logarithmic functions and exponential functions are Inverse of Each other:

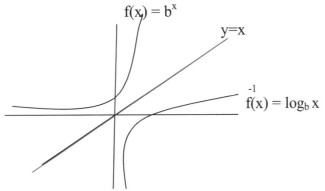

$f(x) = b^x$

y=x

$f^{-1}(x) = \log_b x$

8.2   One – to – One Functions

Definition	A function is one-to-one function if any two different inputs in the domain corresponds to two different outputs in the range, that is if x_1, x_2 are two different inputs of a function f, then $f(x_1) \neq f(x_2)$
Definition of 1-1 function	To test if a function is one-to-one function a horizontal line test is used. If any horizontal line intersects the graph of a function in at most one point, then the function is 1-1 function.
Inverse Functions	Inverse functions are 1-1 functions, the inverse of f(x) is $f^{-1}(x)$ Domain of f(x) = Range of $f^{-1}(x)$ Range of f(x) = Domain of $f^{-1}(x)$
How to find the inverse	To Find the inverse replace x with y, and to find the inverse of a function algebraically, exchange y with x, then solve for y that is the inverse.

Example -1	Determine if the functions are one-to-one functions: a. $f(x) = x^2$ b. $f(x) = x^3$

Solution: a. the graph of the function is a parabola opened up .
if we let $x_1 = -1$, and $x_2 = 1$ Then substituting in gives:
$f(x_1) = f(-1) = (-1)^2 = 1$
$f(x_2) = f(1) = (1)^2 = 1$
Since $f(x) = f(x)$ → then the function $f(x) = x$ is not a 1–1 function as shown.

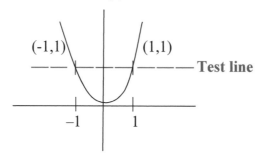

The test line intersects the graph in 2=points, it means not 1–1

b. Choosing the same two points on the domain, and find their $f(x)$:
$f(x_1) = f(-1) = (-1)^3 = -1$
$f(x_2) = f(1) = (1)^{23} = 1$
Since $f(x_1) \neq f(x_2)$ → then the function $f(x) = x^3$ is a 1–1 function.

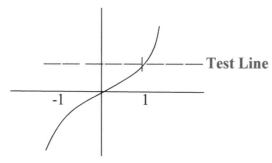

The test line intersects the graph in one point only, it means 1–1

Practice -1	Determine if the graph of the following functions are 1–1:
	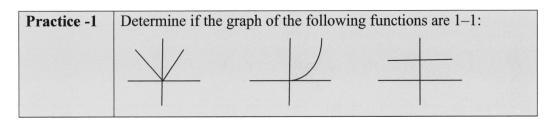

Example -2	Find the in inverse of the following set of ordered pairs: f(x) → { (–2,–8), (–1,–10), (1,1), (2,7), (3,15) }

Solution: we exchange the x-values with y values to get the inverse function.

f(x) → { (–2,–8), (–1,–10), (1,1), (2,7), (3,15)}
$f^{-1}(x)$ → { (–8, –2), (–10, –1), (1,1), (7,2) ,(15,3) }

Obtaining a graph of $f^{-1}(x)$ from graph of f(x):

Example -3	Find the in inverse of the given graph:
	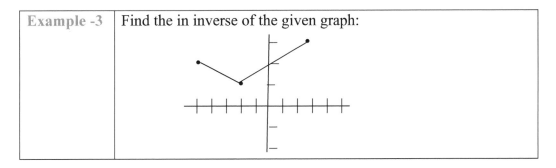

Solution: To find the inverse, we will find the inverse of the points then graph it.

f(x) points	$f^{-1}(x)$ points
–5,2	2,–5
–2, 1	1,–2
3,3	3,3

Then the graph of f(x) and $f^{-1}(x)$ is as follows:

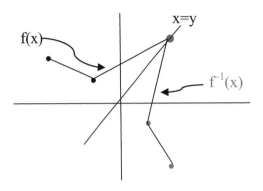

Finding the inverse of f(x) algebraically:

Example -4	Find the inverse of function algebraically: $f(x) = 3x - 1$

Solution: Let f(x) = y

$\quad\quad\quad$ y = 3x −1, then replace y with x

$\quad\quad\quad$ x = 3y −1, now solve for y:

$\quad\quad\quad$ x+1= 3y → then y = 1/3(x+1) = $f^{-1}(x)$

The graph of both f(x) and $f^{-1}(x)$ is as given:

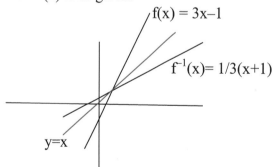

Practice -4	Find the inverse of function algebraically: $f(x) = \dfrac{2x+3}{x+1}$

Practice -5	Find the inverse of function algebraically: $f(x) = x^3 - 8$

Example -6	Find the inverse of function algebraically: $f(x) = (x-2)^2$

Solution: Let $f(x) = y$

$y = (x-1)^2$, then replace y with x

$x = (y-1)^2$, now take the square root of both sides

$\sqrt{x} = y-2$ Solve for y $\rightarrow$ y= $\sqrt{x} + 2 = f^{-1}(x)$

Practice - 6	Find the inverse of function algebraically: $f(x) = \dfrac{x}{4} + 2$

Practice -7	Find the inverse of function algebraically: $f(x) = (x-2)^2$

Convolution	If two functions f(x), and g(x) are inverse of each other then: $f(g(x)) = g(f(x)) = x$

Example -8	Show that the two functions $f(x) = (x+2)^2$, and $g(x) = \sqrt{x} + 2$ are inverse of each other.

Solution: If $f(x)$ is inverse of $g(x)$ then: $f(g(x)) = g(f(x)) = x$

$$f(g(x)) = f(\sqrt{x} + 2) = (\sqrt{x} + 2 - 2)^2$$
$$= (\sqrt{x})^2 = x$$

$$g(f(x)) = g((x-2)^2) = \sqrt{(x-2)^2} + 2$$
$$= x - 2 + 2 = x$$

Practice-6	Show that the two functions $f(x) = 4x{-}8$, and $g(x) = x/4 + 2$ are inverse of each other.

8.3 Exponential Functions

Definition	Exponential functions are of the form: $$f(x) = a^x \quad a \text{ is a positive real number, } x \text{ is a variable}$$ if $a > 1$ → $f(x)$ is an increasing function if $0 < a < 1$ → $f(x)$ is decreasing function
Properties of Exponential functions	• Domain is all the real numbers, range is $y > 0$. • x-axis is the horizontal asymptote • No x-intercepts, and y-intercept is 1 • graph is smooth and continuous Horizontal Asymptote Is the x-axis 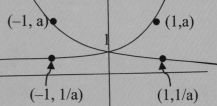 $(-1, a)$ $(1,a)$ $(-1, 1/a)$ $(1,1/a)$

Practice-1	Graph the following exponential functions and check the properties: a. $f(x) = 2^x$ b. $f(x) = 3^x$ c. $f(x) = 4^x$ d. $f(x) = (1/2)^x$ e. $f(x) = (1/4)^x$ f. $f(x) = (1/5)^x$

8.4   Logarithmic Functions

Definition	$y = \log_b x$ IFF $\rightarrow x = a^y$
Properties of Logarithmic functions	• Range is all the real numbers, Domain $x > 0$ • y-axis is the vertical asymptote • No y-intercepts, and x-intercept is 1 • graph is smooth and continuous
Logarithmic functions and Exponential functions	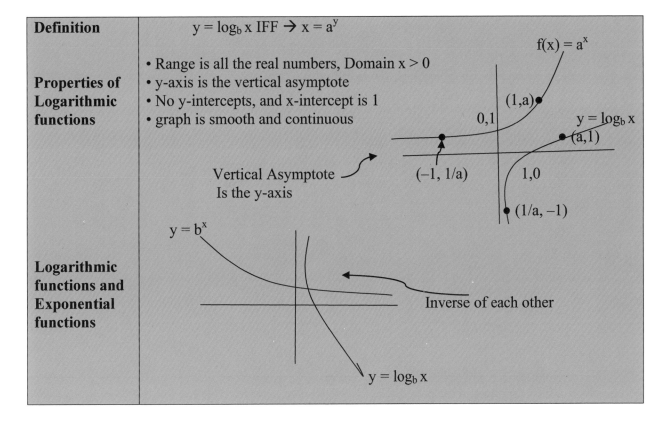

Practice-1	Graph the logarithmic function and check the properties: $f(x) = \ln(x-2)$

Properties of Logarithmic Functions:

$\log_b b = 1$

$a^{\log_a x} = x$

$\log(xy) = \log x + \log y$

$\log_b \dfrac{x}{y} = \log_b x - \log_b y$

$\log_b x^n = n \log_b x$

If $x = y$ → then $\log_b x = \log_b y$

If $a^x = a^y$ → then $x = y$

Change of Base Rule: $\log_b x = \dfrac{\log x}{\log b} = \dfrac{\ln x}{\ln b}$

8.5   Logarithmic and Exponential Functions

Example-1	Solve the exponential equation: a. $5^x = 5^2$ b. $(1/3)^x = 27$

Solution: Using the above rule gives a. → x=2

b. $(1/3)^x = 3^3$

$(1/3)^x = (1/3)^{-3}$ → $x = -3$

Practice - 1	Solve the exponential equation: a. $7^{2x} = 7^2$ b. $(1/4)^x = 32$

Example-2	Solve the exponential equation:
	a. $4^{5x-3} = 16^{x+2}$ b. $e^2 = e^{3x+6}$

Solution: a. $4^{5x-3} = 16^{x+2}$

$$4^{5x-3} = (4)^{2(x+2)}$$

Then, $5x-3 = 2x+4$ → $3x = 7$ → $x = 7/3$

b. $2 = 3x + 6$ → $3x = -4$ → $x = -4/3$

Example-3	Solve the exponential equation: $3^x \cdot 9^{x^2} = 81^x$

Solution: $3^x \cdot 9^{x^2} = 81^x$

$3^x \cdot 3^{2x^2} = 3^{4x}$

$3^{x+2x^2} = 3^{4x}$

$x + 2x^2 = 4x$

$2x^2 - 3x = 0$

$x(2x-3) = 0$ → x=0, or x=3/2 the solution set is { 0, 3/2}

Practice-2	Solve the exponential equation:
	$\dfrac{3^{2x}}{3^{x^2}} = 27$

Solving Logarithmic Equations:

Example-3	Solve the Logarithmic equation:
	$5 \ln_{(2x)} = 30$

Solution: $5 \ln_{(2x)} = 30$

$\ln(2x) = 30/5 = 6 \rightarrow$ but $\ln = \log e$

Then $\rightarrow \log e \,(2x) = 6 \rightarrow 2x = e^6 \rightarrow x = 1/2 \, e^6$

Check: substituting back $\rightarrow$ L.S = $5 \ln (2 \cdot 1/2 \, e^6) = 5 \ln (e^6)$

$= 5.\log_e e^6$

$= 5. \, 6 \log_e e$

$= 30$

$= $ R.S

Practice-3	Solve the following Logarithmic equations:
	a. $\log (5+ x) - \log_6 (x-3) = \log 5$
	b. $\log_3(x+1) + \log_3 (x-5) = 3$

Chapter – 8 Exercise

Determine if the function is one-to-one function:

1. a.   Domain   Range   b.   Domain   Range

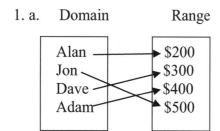
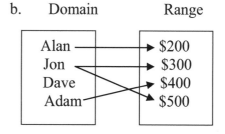

2. {(2,1), (3,3), (4,5), (2,10) }

3. using the horizontal-test-line determine if the graph is one-to-one function:

a.

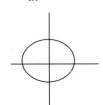

b.

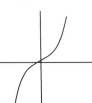

c.

Verify that the functions f, and g are inverse of each other by showing that:
f(g(x)) =g(f(x)) = x

4. $f(x) = \dfrac{3x + 5}{1 - 2x}$ $g(x) = \dfrac{x - 5}{2x + 3}$

5. $f(x) = 5x - 1$ $g(x) = 1/5(x+1)$

6. $f(x) = \sqrt{x + 8}$ $g(x) = x^2 - 8$

7. A graph of function f is given; draw the inverse f^{-1}:

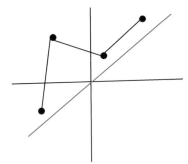

Find the inverse of the following functions algebraically:

8. $f(x) = \dfrac{2x + 1}{x - 1}$

9. $f(x) = \sqrt[3]{x + 4}$

10. $f(x) = \sqrt{x} + 4$

Chapter – 8 Test

1. Determine if the function is one-to-one function:
a. $\{(1, 2), (3, 5), (6, 7), (10, 12)\}$
b. $\{(3, 2), (7, 2), (9, 1), (3, 5)\}$
c. $\{(-2, 1), (0, 2), (3, 4), (5, 4)\}$

2. Use the horizontal-test-line to determine if given function is one-to-one function:

a.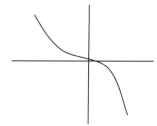

b.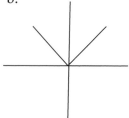

3. Are f, and g inverse of each other?
a) $f(x) = 7x + 1$ $g(x) = x + 1$

b) $f(x) = \sqrt{2x + 4}$ $g(x) = 1/2(x^2 - 4)$

4. Graph the inverse function f^{-1} for the given function f:

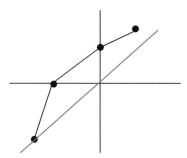

5. Find the inverse function algebraically: $f(x) = \dfrac{3x+1}{2}$

9. Analytic Geometry

9. Analytic Geometry / Conic sections

Objectives: 1. Introduction
2. Distance Formula
3 Parabola
4. Circles
5. Ellipse
6. Hyperbola

9.1 Introduction / Conics

In this chapter we will cover the following: The distance between 2 points and deriving the formula for distance between two points. Conic sections such as circles and the method of completing the square to find the standard formula for circles. Parabola, ellipse and graphs. And hyperbola.

The word **Conic** is derived from word cone, conic sections are curves that results from the intersection of plane with a right circular cone which forms parabola opened to the right or left, a circle, an ellipse or hyperbola.

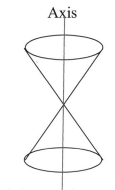

Right Circular Cones

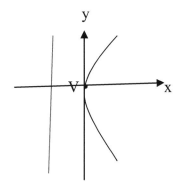

Parabola opened to the right

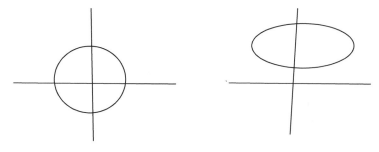

9.2 Distance Formula

The distance d between two points $P_1(x_1, y_1)$, and $P_2(x_2, y_2)$ in the rectangular system can be derived as follows: draw a vertical line at P_2 and a horizontal line at P_1, the two lines will intersect at $P_3(x_2, y_1)$ forming a right triangle. Now we can use Pythagorean Theorem to find the distance d:

Rise $= y_2 - y_1$

Run $= x_2 - x_1$

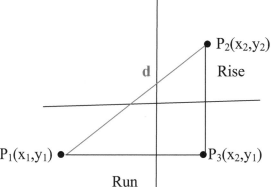

Then by Pythagorean Theorem:

$$d^2 = Run^2 + Rise^2$$

$$d = \sqrt{(x_2 - x_1)^2 + (y_2 - y_1)^2}$$

The mid-point of d is $\left(\dfrac{x_1 + x_2}{2}, \dfrac{y_1 + y_2}{2}\right)$

Example-1	Find the distance between the two points (-2, -5) and (3,7), and find the midpoint between them

Solution: Using the distance formula:

$$d = \sqrt{(3 - (-2))^2 + (7 - (-5))^2} = \sqrt{25 + 144} = \sqrt{169} = 13$$

$$\text{Midpoint} = \left(\frac{-2+3}{2}, \frac{5+7}{2}\right) = (1/2, 1)$$

Practice-1	Find the distance between the two points $(-1/2, -5/4)$, $(1/3, 9/3)$, and find the midpoint between them

9.3 Parabola

In the previous chapter we have discussed the parabola to be a second degree polynomial, and solved problems, here we will concentrate on the vertex of the parabola denoted by v(h, k). Vertex represents the maximum point if the parabola is opened down and minimum point if the parabola is opened up. To find the vertex we will apply the method of completing the square to the second degree polynomial equation or the quadratic equation.

$$f(x) = ax^2 + bx + c$$
$$= a\left(x^2 + b/a\, x\right) + c$$
$$= a\left(x^2 + b/a\, x + (b/2a)^2 - (b/2a)^2\right) + c$$
$$= a\left(x^2 + b/a\, x + (b/2a)^2\right) + c - (b/2a)^2$$
$$= a\left(x + b/2a\right)^2 + (4ac - b^2)/4a$$

$f(x) = (x-h)^2 + k$, where $h = -b/2a$, and $k = f(h) = (4ac-b^2)/4a$

Quadratic equation in Vertex form	If $h = -b/2a$, and $k = (4ac-b^2)/4a$, then $$f(x) = ax^2 + bx + c = a(x-h)^2 + k$$ Where, (h, k) = $(-b/2a, f(/b/2a))$ is the vertex of the parabola, And the axis of symmetry is the line $x = -b/2a$

Max, and Min Of parabola	$V(h,k)$ = max point if $a < 0$ → parabola opened down
	$V(h,k)$ = min if $a > 0$ → parabola opened up
Domain Range	Domain of all parabolas is all the real numbers $d = (-\infty,\infty)$
	Range is $(-\infty, k)$ if parabola is opened down.
	Range is (k, ∞) if parabola is opened up

Example-1	For the quadratic function: $f(x) = 2x^2 + 8x + 5$
	• Find the domain and the range
	• Rewrite the function in vertex form
	• Find the vertex
	• Find the axis of symmetry
	• Graph f(x)
	-

Solution: The Domain is all the real number

Using completing the square method to find the vertex

$$f(x) = 2x^2 + 8x + 5$$

$$(x) = 2(x^2 + 4x) + 5$$

$$= 2(x^2 + 4x + (+ 4/2)^2 + (- 4/2)^2) + 5$$

$$= 2(x^2 + 4x + (4/2)^2) - 2 (4/2)^2 + 5$$

$$f(x) = 2(x^2 + 4x + 4) - 8 + 5$$

$$f(x) = 2(x + 2)^2 - 8 + 5$$

$$f(x) = 2 (x + 2)^2 - 3 \rightarrow \text{this gives } h = - 2, \text{ and } k = - 3$$

The vertex can also be found by using the formula's directly:

a= 2, b=8, and c=5 Then h = –b/2a = – 8/2(2) = –2, and

k = f(h) = f(–2) = 2(–2)^2 + 8(–2) + 5 = 8 – 16 + 5 = –3

V(h,k) = (–2, –3), from this we get the equation of the axis of symmetry line is

h = –2 or x = –2

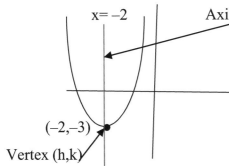

x= –2

Axis of symmetry

Domain = all the real numbers

Range = (k,∞) = (– 3,∞)

(–2,–3)

Vertex (h,k)

Practice-1	For the quadratic function: f(x) = – 3x^2 + 6x + 1 • Find the domain and the range • Rewrite the function in vertex form • Find the vertex • Find the axis of symmetry • Graph f(x)

Example-2	For the quadratic function: f(x) = x^2 – 6x + 9 • Find the domain and the range • Rewrite the function in vertex form • Find the vertex • Find the axis of symmetry • Graph f(x)

Solution: Since $a > 0$ then the parabola is opened up, with minimum point at the vertex (h,k) and the domain of f(x) is all the real numbers and the range = (k,∞), we need to find k.

$$f(x) = x^2 - 6x + 9$$
$$= (x^2 - 6x) + 9$$
$$= (x^2 - 6x + (3)^2 - (3)^2) + 9$$
$$= (x^2 - 6x + (3)^2) - (3)^2 + 9$$
$$f(x) = (x^2 - 6x + (-3)^2) - (-3)^2 + 9 \rightarrow = (x-3)^2 + 0$$

$f(x) = (x-3)^2 \rightarrow$ then h = 3, k=0 this means the parabola touches the x-axis.

Using direct formula will lead to the same solution:

h= -b/2a = $-(-6)/2(1) = 3$, and

$k = f(h) = f(3) = (3)^2 - 6(3) + 9 = 0$

Then axis of symmetry is x=h = 3

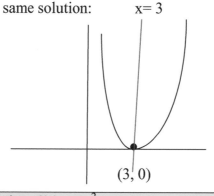

x= 3

(3, 0)

Practice-2	For the quadratic function: $f(x) = 2x^2 + x + 1$ • Find the domain and the range • Rewrite the function in vertex form • Find the vertex • Find the axis of symmetry • Graph f(x)

Example-3	Find the quadratic function from the given information: $V(h,k) = (1,-5)$, and y-intercept $= -3$

Solution: Using the quadratic function in vertex form:

$f(x) = a(x -h)^2 + k$, substitute the values of $h = 1$, and $k = -5$

$f(x) = a(x-1)^2 - 5$ use the y-intercept to find a

Since y-intercept point is $(0, -3)$

Then → $f(0) = -3 = a(x-1)^2 - 5$

$\qquad -3 = a(x^2 - 2x + 1) - 5$

$\qquad -3 = a(0 - 2(0) + 1) - 5$

$\qquad -3 = a - 5$ → $a = 2$

Then the function is $f(x) = 2(x-1)^2 - 5$

Practice-3	Find the quadratic function from the given information: $V(h,k) = (1,-3)$, and y-intercept $= -2$

9.4 Circles

Definition	The circle is defined as a set of points in the xy-plane at a fixed distance r (radius) from the origin if the circle is centered at the origin, or distance r from a point (h,k) if the circle is not centered at the origin

Radius r	The radius r can be found using the distance formula because it is the distance between a point on the circumference and the origin of the circle $r = \sqrt{(x-h)^2 + (y-k)^2}$, (h,k) point at the center If the circle is centered at the origin, then (h,k) =(0,0) and the formula will be: $r = \sqrt{x^2 + y^2}$

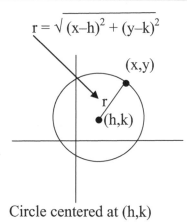

Circle centered at (h,k)

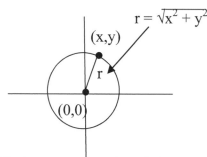

Circle centered at the origin (0,0)

Equation of the circle	Equation of the circle in the standard form: $r^2 = (x-h)^2 + (y-k)^2$ centered at (h,k) And $r^2 = x^2 + y^2$ centered at the origin(0,0)
Equation of the circle in the general form	Equation of the circle in general form is: $x^2 + y^2 + ax + by + c = 0$

Example-1	Find the standard form of the equation of the circle with radius 6 centered at (–3,–5), and graph.

Solution: Given r=6, h= –3, and k= –5, we substitute these values on the standard form:

$$r^2 = (x\text{-}h)^2 + (y\text{ }\text{–}k)^2$$
$$6^2 = (x\text{–}(\text{–}3))^2 + (y\text{–}(\text{–}5))^2$$
$$36 = (x+3)^2 + (y+5)^2$$

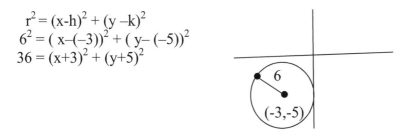

Practice-1	Find the standard form of the equation of the circle with radius 10 centered at (–3, 8), and graph.

Example-2	For the given equation, find the center, the radius and graph the circle: $$x^2 + y^2 + 6x + 4y + 9 = 0$$

Solution: Group x-terms and y-terms separately, and take the constant to the R.S:

$$(x^2 +6x) +(y^2 + 4y) = -9 \text{ complete the square of each group}$$
$$(x^2 + 6x + (6/2)^2 - (6/2)^2) + (y^2 +4y +(4/2)^2 - (4/2)^2) = -9$$
$$(x^2 +6x + 9) + (y^2+4y + 4) = -9 + 9 + 4$$
$$(x+3)^2 + (y+2)^2 = 4$$

Then the circle is centered at (h,k) = (–3, –2) with radius r =2

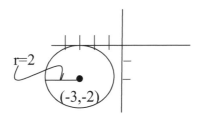

Practice-2	For the given equation, find the center, the radius and graph the circle: $$x^2 + y^2 - 2x - 4y - 4 = 0$$

9.5 Ellipse

The ellipse is a collection of points in the plane. It has two major points called **foci** which are fixed (F_1, F_2), and two axis (major, and minor)

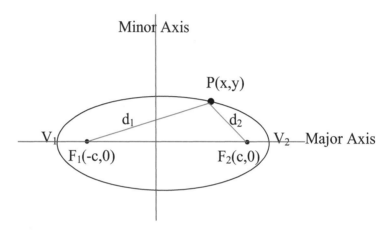

For any point on Ellipse $d_1 + d_2 = \text{constant} = 2a$

Equation of Ellipse With center at the origin	Equations of the ellipse with center at the origin $(0,0)$ and foci at $F_1(-c,0)$ and $F_2(c,0)$ with vertices $V_1(-a,0)$ and $V_2(a,0)$, with the minor axis at $(0,b)$ and $(0,-b)$:
Horizontally	$$\frac{x^2}{a^2} + \frac{y^2}{b^2} = 1$$ where, $a > b > 0$, and $b^2 = a^2 - c^2$

Vertically	$\dfrac{x^2}{b^2} + \dfrac{y^2}{a^2} = 1$ where, $a > b > 0$, and $b^2 = a^2 - c^2$
Equation of Ellipse With center at (h,k)	Equation of the Ellipse at (h,k) is:
Horizontally	$\dfrac{(x-h)^2}{a^2} + \dfrac{(y-k)^2}{b^2} = 1$
Vertically	$\dfrac{(x-h)^2}{b^2} + \dfrac{(y-k)^2}{a^2} = 1$

Finding the equation of Ellipse

Applying the formula: $d_1 + d_2 = 2a$, then using distance formula gives,

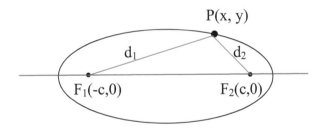

$d_1 + d_2 = 2a$

$d_1 = \sqrt{(x+c)^2 + y^2}$, and $d_2 = \sqrt{(x-c)^2 + y^2}$, then,

$\sqrt{(x+c)^2 + y^2} + \sqrt{(x-c)^2 + y^2} = 2a$

$$\sqrt{(x+c)^2 + y^2} = 2a - \sqrt{(x-c)^2 + y^2} \quad \text{Squaring both sides,}$$

$$(\sqrt{(x+c)^2 + y^2})^2 = (2a - \sqrt{(x-c)^2 + y^2})^2$$

$$(x+c)^2 + y^2 = 4a^2 + (x-c)^2 + y^2 - 4a\sqrt{(x-c)^2 + y^2}$$

$$x^2 + 2xc + c^2 + y^2 = 4a^2 + x^2 - 2xc + c^2 + y^2 - 4a\sqrt{(x-c)^2 + y^2}$$

$$4a\sqrt{(x-c)^2 + y^2} = 4a^2 - 4xc \quad \text{Dividing by 4 and squaring both sides,}$$

$$(a\sqrt{(x-c)^2 + y^2})^2 = (a^2 - xc)^2$$

$$a^2(x^2 - 2xc + c^2 + y^2) = a^4 - 2a^2xc + (xc)^2$$

$$a^2x^2 - 2a^2xc + a^2c^2 + a^2y^2 = a^4 - 2a^2xc + x^2c^2$$

$$a^2x^2 - x^2c^2 + a^2y^2 = a^4 - a^2c^2$$

$$x^2(a^2 - c^2) + a^2y^2 = a^2(a^2 - c^2) \quad \text{But } b^2 = a^2 - c^2$$

$$b^2x^2 + a^2y^2 = a^2b^2 \quad \text{dividing both sides by } a^2b^2 \text{ Gives,}$$

$$\frac{x^2}{a^2} + \frac{y^2}{b^2} = 1 \quad \text{Equation of the ellipse at the center in a horizontal position}$$

Example-1	Find the equation of ellipse with one focus at (2, 0), and the vertex at (-3, 0).

Solution: from the given information we get:

$$c = 2, a = 3 \rightarrow \text{then } b^2 = (3)^2 - (2)^2 = 5$$

The equation of the ellipse is: $\dfrac{x^2}{9} + \dfrac{y^2}{5} = 1$

Practice-1	Find the equation of ellipse with one focus at (4, 0), and the vertex at (– 6, 0), and graph the equation.

Example-2	Find the equation of ellipse with center at (2, –2), vertex at (7,–2) and focus at (4,–2).

Solution: from the given information we get:

Here the ellipse is not centered at the origin but at (h,k)= (2,–2). Since all center, vertex, and focus have y= –2, then the major axis is y= –2 or parallel to x-axis.

The distance from center (2,–2) to focus (4,–2) = 4 –2 = 2 → or c=2.

The distance from the center (2,–2) to vertex (7,–2) = 7–2 = 5 → or a=5

Then $b^2 = a^2 - c^2 = 25 - 4 = 21$, now using the formula for the horizontal ellipse:

$$\dfrac{(x-h)^2}{a^2} + \dfrac{(y-k)^2}{b^2} = 1$$ we get the equation of the ellipse as:

$$\dfrac{(x-2)^2}{25} + \dfrac{(y+2)^2}{21} = 1$$ we get the equation of the ellipse as:

Practice-2	Find the equation of ellipse with center at (–3, 1), vertex at (–3, 3) and focus at (–3, 0). Graph the equation.

9.6 Hyperbola

The ellipse is a collection of points in the plane. It has two major points called **foci** which are fixed (F_1, F_2), and two axis (Transverse, and Conjugate).

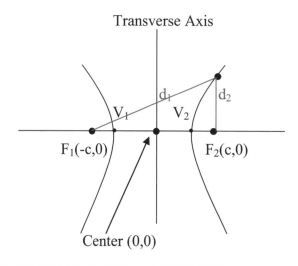

For any point on Hyperbola $|d_1 - d_2|$ = constant = 2a

Equation of Hyperbola With center at the origin	Equations of the Hyperbola with center at the origin $(0,0)$ and foci at $F_1(-c,0)$ and $F_2(c,0)$ with vertices $V_1(-a,0)$ and $V_2(a,0)$, with the minor axis at $(0,b)$ and $(0,-b)$:
Horizontally	$\dfrac{x^2}{a^2} - \dfrac{y^2}{b^2} = 1$ where, $a > b > 0$, and $b^2 = c^2 - a^2$
Vertically	$\dfrac{y^2}{a^2} - \dfrac{x^2}{b^2} = 1$ where, $a > b > 0$, and $b^2 = c^2 - a^2$

Intermediate Algebra with Analytic Geometry

Equation of Hyperbola With center at (h,k)	Equation of the Hyperbola at (h,k) is:
Horizontally	$$\dfrac{(x-h)^2}{a^2} - \dfrac{(y-k)^2}{b^2} = 1$$
Vertically	$$\dfrac{(y-h)^2}{a^2} - \dfrac{(x-k)^2}{b^2} = 1$$

Example-1	Find the equation of hyperbola with center at the origin, one focus at (4, 0), and one vertex at (–1, 0).

Solution: From the given information we have:

One focus is at (3, 0) = (c, 0) → c =3

One vertex is at (–1, 0) = (–a, 0) → a = 1

Then → $b^2 = c^2 - a^2 = 9 - 1 = 8$

Equation of the hyperbola is:

$$\dfrac{x^2}{1} - \dfrac{y^2}{8} = 1$$

Practice-1	Find the equation of hyperbola with center at (0, 0), vertex at (1, 0) and focus at (3, 0). Graph the equation.

Example-2	Find the equation of hyperbola with center at (2, –2), vertex at (4,–2) and focus at (5,–2).

Solution: Here the hyperbola is not centered at the origin but at (h,k) = (2,–2)

y= –2 for the center, focus, and vertex this means they lie on axis that is Parallel to x-axis.

The distance from center (2,–2) to the focus (5,–2) = 5–2 = 3 → c=3
The distance from center (2,–2) to the Vertex (4,–2) = 4–2 = 2 → a=2
Then $b^2 = c^2 - a^2 = 9 - 4 = 5$.
→ Equation of the hyperbola is:

$$\frac{(x-2)^2}{4} - \frac{(y+2)^2}{5} = 1$$

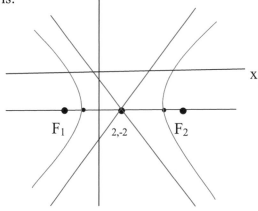

Practice-2	Find the equation of hyperbola with center at (1, 3), vertex at (0, 3) and focus at (–2, 3). Graph the equation.

Chapter – 9 Exercise

Find the distance between each pair of points and find the mid points:
1. (2,5, 1.2) and (1.5, – 5.4)
2. $(-\frac{1}{3}, -\frac{1}{6})$ and $(\frac{3}{4}, \frac{5}{6})$
3. $(2\sqrt{2}, 5\sqrt{6})$ and $(-5\sqrt{2}, 4\sqrt{6})$

For the given center and radius of the circle, write the equation in the standard form:
4. Center at (0, 0) and r=5
5. Center at (–3, 7) and r= 4
6. Center at (– 7, -2) and r = $\sqrt{3}$

Give the center and the radius of the circle described by the given equation:

7. $x^2 + y^2 = 4$

8. $(x-2)^2 + (y-3)^2 = 1$

9 $(x-5)^2 + (y+4)^2 = 16$

10. Complete the square and write the equation in the standard form:
$$x^2 + y^2 + 12x - 6y = 4$$

11. Graph the ellipse: $\dfrac{(x+2)^2}{16} + (y-2)^{2} = 1$

12. Graph the ellipse: $4(x-1)^2 + (y-3)^2 = 4$

Chapter – 9 Test

1. Find the distance between the two points and give their midpoint:
   $(-7/2, 10/3)$ and $(-11/2, -7/3)$

2. For the given center and radius of a circle, find its equation in the standard form:
   Center at $(-3, 6)$ and radius $= 6$

3. Find the center and the radius of the given equation:
   $(x+3)^2 + (y+6)^2 = 9$

4. Complete the squares to find the center and radius of the circle:
   $x^2 + y^2 = 2x + 15$

5. Graph the ellipse: $\dfrac{(x+2)^2}{16} + (y-2)^{2} = 1$

6. Graph the ellipse: $9(x-1)^2 + 4(y-3)^2 = 36$

7. Graph the hyperbola: $9x^2 = 9 - y^2$

Answers to Exercises

Chapter-1

1. 6   2. –20   3. – 13   4. – 5   5. 6   6. 72/11   7. –27/4   8. – 6/5

9. Inconsistent   10. Identity   11. Conditional   12. Inconsistent   13. {20, 40, 120}

14. a) \$6500 b) 1750 watches   15. 400000000   16. 0.0000005   17. 18 ft^2

18. 60 ft^3 19. Smaller number = 10, larger number = 16   20. {9, 11}

Chapter-2

1. {9, 1}   2. {1, 2}   3. {5, – 1}   4. {0, –1}   5. No Solution

6. {–3/2, 11/4, 19/4}   7. {– 1, 2, – 2}   8. {–4, 12, 0}   9. {3/2, 0, ½}

10. {–2, 3, 5}   11. Equations: x+y = 170; 0.1x + 0.2 y = 27.2 → x=68, y=102

12. The system of equations is:
$$x+ \quad y+ \quad z=10{,}000$$
$$0.5x + 0.07y + 0.10z = 695$$
$$x \qquad\qquad – z = 1500$$
Solution set is {\$4000, \$3500, \$2500}

13. Let x= rate of the plane
        y=rate of the wind
Equations are: 1.5 x +1.5y = 400
                3x – 3y = 400
The solution set is {200 mile/hr, 66.7 mile/hr}

Chapter-3

1. x < 3 $\longleftarrow$ $)$
   3

2. y ≤ 3/2 x + 3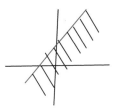

3. y > x/3 + 20/3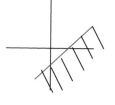

5. y ≥ –3/4 x + 3/4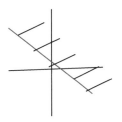

4. y ≤ x–3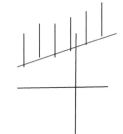

6. y ≤ –3x

7. 1. y ≥ –2x + 2
   2. y ≤ x – 5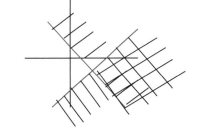

8. 1. y ≥ – 3/2 x – 7/2
   2. y ≤ – x – 9

Solution region

9. a. x > 3

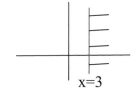

x=3

b. x < −1

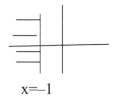

x=−1

c. y ≥ 3

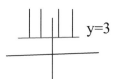

y=3

c. y ≤ 10

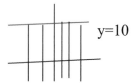

y=10

9.

Points	Maximum z = x + 2y
0,0	z = 0
0,3	z=6 Max value at 0,3
3,0	z=3

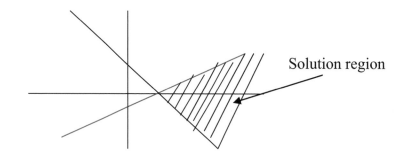

Solution region

Chapter-4

1. $4(x-2)$ 2. $3x(x-3y+4y^2)$ 3. $(x-9)(x+9)$ 4. $(7x-3)(7x+3)$

5. $(12x-6y)(12x+6y)$ 6. $(2x+3)(4x^2-6x+9)$ 7. $(4+5x)(16-20x+25x^2)$

8. $(6x-3)(36x^2+18x+9)$ 9. $(2-3x)(4+6x+9x^2)$

10. $3x(2x+3)+2(2x+3)=(2x+3)(3+2)$ 11. $2x(x-2)-(x-2)=(x-2)(2x-2)$

12. $(3x-4)(3x-4)$ 13. $-(16x^2+11x-5)=(5-16x)(x+1)$

14. Quotient $= 3x^2+11x+32$, remainder $= 99$ 15. Quotient $= 2x^3+2$, remainder $= 0$

16. Synthetic Division gives remainder $r = -4$, and $f(-2) = -4$

17. Synthetic division gives remainder $r=23$, and $f(2) = 23$

18. $(x-6)(x-6)=0$ Solution set is → {6}   19. $(x-6)(x-4)=0$. Solution set is → {4, 6}

20 Solution set is {−3, 5/2}   21. Solution set is: {(−∝, −1) ∪ (3,∝)}

22. Solution set is {[2, 4]}   23. Solution set is {[4, ∝)}

24. a) 120 feet, b) $t = 1.38s$, or $t = 3.62$ s (use quadratic formula)

25. Width = 12 yard, Length = 15 yards, Fence = 54 yards.

Chapter-5

1. $\dfrac{x+2}{x+6}$ 2. $\dfrac{2x(x+2)}{(x^2+2x+4)}$ 3. $\dfrac{x+3}{x-2}$ 4. $\dfrac{x}{3}$ 5. $\dfrac{x-5}{2}$

6. $\dfrac{1}{x-2y}$ 7. $\dfrac{5}{3x}$ 8. $\dfrac{x+3}{x+1}$ 9. $\dfrac{7x-2}{(x+1)(x-2)}$ 10. $\dfrac{x+1}{7x-4}$

11. $\dfrac{x+1}{x-2}$ 12. $\dfrac{-5(x-8)}{(x-5)(x+5)}$ 13. x= –24   14. x = 10/3   15. { –3, 2}

16. (–2 , 4]   17. { (–∝, –1) ∪ [1,2]}   18. {(– 2, 2) ∪ (10, ∝)}   19. $ 360

20. x = 120 bicycles.

Chapter-6

1. a) 3/4   b) 3   c)13   2. x ≥ 3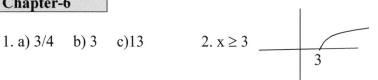

3. a) 3/10   b)–1   c) y   d) 1/10   e) (x+5)   4.Domain is [2, 7)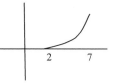

5. x=22   6. {3, 7}   7. {0, 3}   8. x=5/2   9. No solution   10. x=44

Chapter-7

1. {± 7}   2. {–4,2}   3. {–2, 2/3}   4. {–3 ± √5}   5. {1/4(5± 3√10)}

6. {1/2(3± √29)}   7. {–1 ± √6 /2}   8. { 1/10 (3± √54)}   9. {1/4(–1 ± √57)}

10. {1/2(3 ± √21)}

Chapter-8

1. a) yes   b) No   2. No   3. a) No   b) yes   c) yes   4. Yes   5. Yes

6. yes   8. $f^{-1} = \dfrac{x+1}{x-2}$ 9. $f^{-1} = x^3 - 4$ 10. $f^{-1} = (x-4)^2$

Chapter-9

1. Distance ≈ 6.68 2. Distance ≈ 1.47 3. Distance ≈ 10.2

4. $x^2 + y^2 = 25$ 5. $(x+3)^2 + (y-7)^2 = 16$ 6. $(x+7)^2 + (y+2)^2 = 3$

7. C=(0,0); r=2   8. C=(2,3); r=1   9. C= (5,2); r=4   10. $(x-6)^2 + (y-3)^2 = 49$

11. (h,k) =(–2,2), a =4, b=1, v_1 = (–6, 2), v_2 = (2,2)

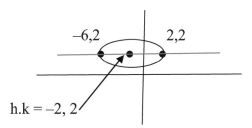

12. (h,k) =(1,3), a =2, b=1, v_1 = (1, 5), v_2 = (1,1)

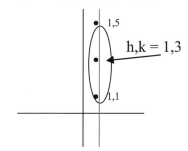

INDEX

V

CPSIA information can be obtained at www.ICGtesting.com
Printed in the USA
BVIW12n0012170818
524650BV00004B/13